Wittgenstein Flies a Kite

A Story of Models of Wings and Models of the World

Susan G. Sterrett

Pi Press
New York

PI PRESS

An imprint of Pearson Education, Inc.
1185 Avenue of the Americas, New York, New York 10036

© 2006 by Susan G. Sterrett

Printed in the United States of America

First Printing: November 2005

Library of Congress Cataloging-in-Publication Data
Sterrett, Susan G.
  Wittgenstein flies a kite : a story of models of wings and models of
the world / Susan G. Sterrett.
    p. cm.
  Includes bibliographical references and index.
  ISBN 0-13-149997-1
  1. Wittgenstein, Ludwig, 1889-1951. 2. Wittgenstein, Ludwig, 1889-
1951. Tractatus logico-philosophicus.   I. Title.
  B3376.W564S8765 2005
  192—dc22

                                                      2005023489

Pi Press books are listed at www.pipress.net.

ISBN 0-13-149997-1

Pearson Education LTD.
Pearson Education Australia PTY, Limited
Pearson Education Singapore, Pte. Ltd.
Pearson Education North Asia Ltd.
Pearson Education Canada, Ltd.
Pearson Educación de Mexico, S.A. de C.V.
Pearson Education—Japan
Pearson Education Malaysia, Pte. Ltd.

Contents

Preface

Stories about thinkers who have solved problems once thought insoluble string together three key moments. First, there is the private moment in which the thinker glimpses the possibility of a solution. Later, a moment in which the solution is actually worked out, validating that first glimpse. And, finally, a crowning moment, in which the wider world recognizes that the problem once thought insoluble has really been solved. The story of a great accomplishment then turns to its subsequent consequences, and so fans out into numerous threads, the strands of each thread eventually tapering into the fabric of history, becoming part of the background against which later moments in other people's lives are lived.

This book traces threads in the background of the first private moment of just such a story: the moment of insight in which a young aeronautical researcher-turned-philosopher, Ludwig Wittgenstein, glimpsed the "fundamental thought" (or *Grundgedanke*, in his native German) of his first book, *Tractatus Logico-Philosophicus*. At the time he completed the manuscript for his book, he was confident it contained the solution to all the problems of philosophy. It would become one of the most well-known philosophical works of the twentieth century.

Some of the threads in the background of his first private moment of insight were the indirect consequences of the solution to another, completely different kind of problem. That problem, too, dated from ancient times, and its solution had been unsuccessfully attempted so many times that it had been deemed insoluble. That earlier story is the story of the invention of the airplane, culminating in the Wright Brothers' discovery of the solution to the problem of controlling powered human flight. Thus, the solution to the sophisticated technical problem of controlled flight seeded the later solution of a foundational problem in logic and the philosophy of language.

That is the theme of this book: to show the connection between two disparate inventions of the twentieth century. One is an advance in science and technology—the Wright Brothers' invention of the airplane; the other, an advance in the foundations of logic and language—Wittgenstein's account of the world as consisting of "facts, not things," and his account of language on which a proposition is a "model of reality," as first presented in his *Tractatus Logico-Philosophicus*. In the story I tell here, the connection between these two stories is not the sort of connection in intellectual history that was necessarily bound to occur, so that if it had not come about in the particular way it had, it would have come about in another. Rather, my story goes, it is something of a fluke that this connection between two disparate disciplines ever occurred at all. It just so happened that one individual physicist with a peculiar mix of interests happened to become interested in the question of the relationship between empirical equations and models at a certain moment in time, and that it was due to the invention of the airplane.

It is also something of a fluke that I came to write this book. Due to a chance set of circumstances, I happened to be referring to a paper on the methodology of scale models, Edgar Buckingham's "On Physically Similar Systems," at the same time I was rereading the *Tractatus*, and I noticed some striking thematic similarities between the *Tractatus* and that paper. There were also a few details, peculiar ways of putting things, that the papers had in common, suggesting some cultural commonalities. Could Wittgenstein have read this paper, possibly while in engineering research? After exhaustive searches, I found that no historian or philosopher—nor anyone at all—had ever so much as hinted at any connection between the two thinkers or their work. In fact, I found that Buckingham's paper on similar systems only appeared in 1914 and had not even been written when Wittgenstein was a student in aeronautical research, for he left Manchester for Cambridge in 1911.

After this initial discouragement, I put the question aside as a sort of intellectual misfire. Then I read the memoir of Wittgenstein by his friend G. H. von Wright, according to which Wittgenstein had on numerous occasions explained the importance that thinking about scale models had to his having a crucial moment of insight in 1914:

There is a story of how the idea of language as a picture of reality occurred to Wittgenstein. [There exist several somewhat different versions of it.] It was in the autumn of 1914, on the eastern front. Wittgenstein was reading in a magazine about a lawsuit in Paris concerning an automobile accident. At the trial a miniature model of the accident was presented before the court. The model here served as a proposition; that is, as a description of a possible state of affairs. It has this function owing to a correspondence between the parts of the model (the miniature-houses, -cars, -people) and things (houses, cars, people) in reality. It now occurred to Wittgenstein that one might reverse the analogy and say that a proposition serves as a model or picture, by virtue of a similar correspondence between its parts and the world. The way in which the parts of the proposition are combined—the structure of the proposition—depicts a possible combination of elements in reality, a possible state of affairs.

My initial question was now too tantalizing to ignore. Though I am a philosopher and not a historian, I decided to investigate whether there might be a connection between Buckingham's "On Physically Similar Systems" and Wittgenstein's insight. After all, Wittgenstein spoke of a proposition being a fact that pictures a fact—just as physical things model other physical things. Thus, the notion of physical similarity might be the notion he had drawn upon.

Historical studies that existed then looked for scientific and technological influences on Wittgenstein in his education and the intellectual surroundings in his youth. As I began to read up on the territory previously covered, I found that there was at that time an established list of suggestions around which the historical work regarding technological influences clustered:

Dynamic models in Heinrich Hertz's *Principles of Mechanics*

Descriptive geometry learned in the Technische Hochschule

Franz Reuleaux's models of machine components

Manchester Professor Horace Lamb's *Hydrodynamics*

Experimental work on jet spray through nozzles

Wittgenstein's own patented propeller design

These suggestions, however, were all elements in his milieu available during the years *prior to* his arrival in Cambridge. Hence, it occurred to me that these studies probably had not considered currents in scientific and technological thought that occurred only after he had arrived at Cambridge to study logic and philosophy—from late 1911 on. I was encouraged to pursue my project further when I found that, although Buckingham was a generation ahead of Wittgenstein, when I constructed a timeline laying out events of their lives side by side, it revealed that Buckingham's first paper on the subject of similar systems appeared just a few months before Wittgenstein had his insightful moment occasioned by thinking about scale models. Upon further reading, some other unexpected connections showed up, such as the fact that Wilhelm Ostwald played a role both in Buckingham's education and in the publication of the *Tractatus*: Buckingham had studied for his doctorate under Ostwald, and Ostwald was the philosopher-scientist who had taken an interest in publishing Wittgenstein's manuscript when so many others were uninterested. Then I noticed that there were a variety of publications on the topic of similarity in England around 1914. More and more elements of an unanticipated landscape began to come into view.

In this new conceptual landscape, the notions of "object," "fact," and "state of affairs" in Wittgenstein's *Tractatus* took on a wholly new meaning, and a somewhat coherent view clicked into place. In this new view, a number of previously perplexing statements in the *Tractatus* became clear. When I realized that what Wittgenstein called "the fundamental thought" of the *Tractatus*—that there are no logical constants—had a very close analogue in statements about "the most general form of an empirical equation" in Buckingham's paper, the idea for this book was born. I presented a sketch of the project in Vienna, Austria in the summer of 2000 at the History of Philosophy of Science meeting and, almost five years later, the manuscript that became the book you hold finally made it to my publisher.

The story in this book is thus about how a philosophical insight that led to one of the major philosophical movements of the twentieth century came about in part due to technical work attendant upon the political consequences of a technological advance. The paper "On Physically

Similar Systems" was written by an American physicist trained in Germany, but the direction of his attention to the question of physically similar situations was occasioned by political events that made national competence in flight technology a priority. The technological advance whose consequences occasioned this fusion of ideas was the use of heavier-than-air military aircraft, which in turn was a result of the success of the Wright Brothers' endeavors to build a flying machine that was capable of sustained, controlled, manned flight.

Looking back to when the Wright Brothers began their research takes us to a point in time almost 15 years prior to Wittgenstein's crucial moment of insight, to just before the turn of the century. It was around 1900, when Wittgenstein was about ten years old, that Wilbur Wright had the moment of insight in which he felt he had the solution to control of manned heavier-than-air flying machines. This was the now-legendary moment when he twisted a cardboard box in which bicycle inner tubes were shipped in his hands, and suddenly saw in it a means of controlling glider wings. He explained it to his brother Orville, and, after a sleepless night of contemplation, he too became convinced that it would make practical manned heavier-than-air powered flight possible. Whether or not anyone else had ever thought of wing warping, to them it was new, and they recognized its significance for their problem. It was the start of the path that would very rapidly lead to the invention of a practical flying machine.

The Wright Brothers had been interested in flight since their boy-hoods, when their father had brought them a helicopter-style toy called a "bat" that was wildly popular in Europe. It was an amazing toy, able to stay aloft for appreciable periods. It was based on a design by Alphonse Pénaud, an aeronautical researcher in France who designed his flying machine models to be inherently stable. The Wrights had for years tried to make bigger versions of that helicopter toy, but bigger versions didn't fly, and their technological adventurousness was directed to other pursuits—first printing presses, and then optimizing the newer-style, high-precision "safety" bicycles. They never lost interest in the helicopter toy, though. Their nephews reported that they were still constructing different versions of it many years later, and Wilbur himself said they had spent many years constructing replicas of various sizes.

Otto Lilienthal's death in 1895 occasioned obituaries and news stories of his accomplishments, especially his manned gliding experiments. Reading of his feats in the newspaper renewed the Wright Brothers' interest. They experimented with kites and gliders, trying different materials and methods of control, yet still making little progress. It was not until that first moment of insight a few years later, when Wilbur twisted the cardboard bicycle tire inner-tube box, that they became convinced they could solve the problem of flight. By then, they had established a business sufficiently successful to support their research into flight. They had a specific vision to follow, and they had their own means of support. They set about experimenting to validate their vision.

The next step was to gather everything they could read on what others had learned so far. By that time, there was a great deal of information available, including the data that Lilienthal had collected from his experiments in Berlin, and a compendium of information titled "Progress in Flying Machines" collected by Octave Chanute, a retired Paris-born American engineer who had begun performing his own experiments in flight. Chanute had collected information on the different kinds of flying machines that had been designed, experiments done with them, and how they had performed. The Wrights used the data to design the shape and size of the wing, and then located a place that would provide sufficient winds in which to carry out their experiments.

Now, to test their ideas. They were aware that many had lost their lives in pursuit of the same entrancing goal; not only Lilienthal, but Percy Pilcher, too, had recently died in an attempt to fly in a machine of his own design. The Wright Brothers first flew the gliders only as kites, to make sure that their idea about control would work before designing and building man-carrying ones. But their gliders didn't get as much lift as their calculations had led them to expect. What was at fault—the machine or their expectations? Their experiments with a wind tunnel made of cardboard boxes and a fan, and a cleverly designed apparatus to hold the mini-wing surfaces, were decisive; it was their expectations that were at fault. They needed better experimental data about lift on wing surfaces. They decided to generate it themselves. They built a larger, better wind tunnel, and systematically generated their own data on a wide range of shapes and wind speeds, which they used to calculate lift forces and redesign the wings of their flying machine. They generated all over again the same data that had been collected by Lilienthal. They would later learn that they had not been properly informed about how to read Lilienthal's tables, but the important thing was that the data they produced with a wind tunnel of their own design was accurate and enabled them to properly design the wings of their gliders. Their initial conviction that they had a solution was validated. Their proof that their solution worked lay in the flights made by their flying machine.

The rest is history: after perfecting the design of a glider and controls for it, and developing an engine to power it, they did succeed, closed down their experiments for awhile, and waited for their patent to be approved. The story has twists and turns, but their success became fully recognized by the world at large in 1908, in heavily attended public exhibitions held in Europe and the U.S. Ironically, this final crowning moment occurred just after Wittgenstein left the Continent for England to begin a career in aeronautical research; he spent that summer in the North of England designing, building, and flying kites at an experimental kite-flying station.

Airplanes were soon considered necessary elements of a modern nation's arsenal. Wilbur Wright's public demonstrations in late 1908 were soon followed by Bleriot's flight across the channel separating England from continental Europe in 1909, which had swift and significant consequences as well as symbolic significance. England initiated an Advisory Committee on Aeronautics that same year to coordinate and support research related to progress in aeronautics, and it built wind tunnel research facilities. There was then no question that the problem of flight had been solved; the question was only how rapidly each country could progress in its aeronautical capability. In 1911, Wittgenstein rather hastily left his position as a research student in engineering at Manchester to study logic and philosophy with Bertrand Russell in Cambridge. By then, several European countries had major aeronautical research facilities and state-sponsored agencies to support and direct them as well. The U.S., on the other hand, had no government agency devoted to aeronautical research when the prospect of war in Europe loomed in 1913.

The political situation in Europe created a sense of urgency in the U.S. Many within the U.S. administration were becoming increasingly concerned about the country's lack of initiative in developing aeronautical research capabilities, and they wanted facilities that at least kept up with those in Europe. Congress was urged to develop similar research facilities in the U.S. The importance of U.S. wind tunnel facilities was brought to the attention of a number of scientists in government agencies, among them Edgar Buckingham, who worked at the U.S. National Bureau of Standards (NBS). The NBS set standards for measurements used in research and so would be involved in instrumentation used in wind tunnels. Buckingham, though relatively unknown to the outside world, was distinguished in his profession. As did many New Englanders of his class and interests, he had traveled to Germany for advanced study—in his case, for a doctorate in physics. He had investigated fluorescence with Ostwald in Leipzig, then had written a book on the foundations of thermodynamics. After a few years spent in teaching and research at American colleges and universities, he settled into a government career as a physicist, investigating topics deemed relevant to the nation's needs. The special set of circumstances that led to a paper entitled "On Physically Similar Systems" arose sometime after 1911, when he first became involved in providing advice on something with which he had little previous involvement: laboratory research into heavier-than-air flight.

Thus, around 1912, Buckingham set out to identify the foundations of the methodology of using experimental models for aeronautical research. He wanted to know: what were the assumptions made in the process, and what followed as a matter of logic? The practical methods themselves were nothing new; the U.S. Navy rightfully claimed that it was already doing very similar work in its laboratories devoted to research on ships in canals and on performance of screw propellers, and that testing in a wind tunnel would not be a qualitatively different task. Buckingham, though, was a philosophically minded physicist, always scrutinizing foundations, and this meant first clarifying the foundations of the practice of engineering scale models, whether in water canals or wind tunnels. What he sought to do in "On Physically Similar Systems," which appeared first in July 1914 and then in a greatly expanded form in October 1914, went far beyond clarifying the current practice. He showed that the method could be made more general, so that it applied to anything that could be described in the language of physics, and, further, that the principle providing the foundation of the method was a principle of logic. In an unlikely mix of logic, engineering, and thought experiment of a physics paper, he applied the principle both to the theoretical question of whether we would be able to detect if the universe had shrunk in size overnight to a miniature of itself, and to the question of how to shrink a submerged screw propeller to a miniature situation such that it performed similarly to the original situation.

A number of other papers and lectures on similarity appeared all at once in the year before the war, some in engineering and some in thermo-dynamics. Buckingham's was unusual in its generality and in its emphasis on symbolism. Perhaps only someone with his particular experience could have conceived the sort of paper that Buckingham wrote. He brought the kinds of considerations discussed among philosophers and scientists in intellectual circles in Vienna and German-speaking Europe to questions that had arisen in quite a different kind of community—the community of advocates and researchers of flight. At that time, the parties gathered around the problems of flight formed a motley collection—industrialist-inventors and engineers, physicists and mathematicians, daredevils and mechanical geniuses. They had various degrees of knowledge and employed various degrees of rigor. What brought them into contact was a common, often consuming, desire to pin down the principles that could explain why kites and birds, though heavier than air, could stay aloft in it. It would be no accident that the philosopher who picked up on the import of the philosophical aspects of Buckingham's paper and worked out the implications of what he had said about scientific equations for the more general case of human language in general was also a rare sort of person who had spent time in both these worlds.

That philosopher was Ludwig Wittgenstein. He too was familiar with both these disparate worlds. His intimate acquaintance with the technical issues in heavier-than-air flight came in part from his time at an experi-mental research station in Glossop, in the north of England, designing, building, and flying kites. He was also familiar with the impassioned arguments in Europe about the role of pictures, models, and equations in the foundations of physics, especially the debates about the role of mod-els and equations in scientific theories ignited by the birth of statistical thermodynamics—the same debates and debaters that Buckingham had encountered studying with Ostwald in Germany. We know that

Wittgenstein was very interested in these debates as a young man, that he had read and admired Boltzmann's *Popular Writings* and Hertz's *Principles of Mechanics Presented in a New Form*, and that he said his thinking had been influenced by both of these physicists' works.

But Wittgenstein brought even more to this fusion of ideas. Wittgenstein had gone on to inhabit yet another world besides the two Edgar Buckingham had encountered prior to probing the foundations of experimental models and the theory of dimensions. He had hungrily sought to learn about logic and philosophy, and he learned the terrain of that world from the two people who were rearranging it to provide foundations for mathematics: the German mathematician Gottlob Frege and the renowned Cambridge philosopher Bertrand Russell. Russell, too, saw his own work as a matter of solving problems. Russell felt that in his major work on the foundations of mathematics, *Principia Mathematica*, he had solved the problems responsible for the existence of age-old paradoxes in previous logics. A crucial part of this solution was a set of rules about "types" of variables. This set of rules, encapsulated in what he called a "theory of types," added restrictions in logic that were meant to prevent vicious circles—and, in turn, the ability to formulate the logical paradoxes that had been the subject of controversy since ancient times. Russell considered Wittgenstein a uniquely gifted student and his intellectual colleague and heir, but Wittgenstein thought Russell's solution had problems of its own. He thought Russell's restrictions a desperate and unprincipled move. At first, Wittgenstein was not sure what a better solution might be; he was just certain that Russell's way out was not a principled means of avoiding the paradoxes. It was these sorts of problems—problems in the landscape of logic and foundations of mathematics limned by Frege and Russell—that were occupying Wittgenstein when, in 1914, Buckingham presented the striking characterization of the foundations of the methodology of scale modeling in "On Physically Similar Systems."

One of these problems in logic and philosophy is the problem at the heart of the story in this book: roughly, the problem of how a proposition or statement represents something in the world. The first private moment in the story of its solution is that crucial moment of insight already mentioned. It took place in September 1914. Wittgenstein was at an unusual point in his life. His father, an imposing figure in his country, his family, and Ludwig's own life, had passed away the previous year. After finishing the school year at Cambridge, where he was enrolled as an undergraduate, Wittgenstein left without completing the program in which he was enrolled. He spent a year in a small cottage in Norway. During that year, spent in (almost total) isolation, he produced a manuscript on logic. He emerged to find Europe in crisis. When war was declared in August 1914, he volunteered to serve in the Austrian Army.

According to his sister's memoirs, Wittgenstein at that time "had an intense desire to take on something difficult and demanding and to do something other than purely intellectual work." And so the first few months of World War I find him once again in isolation of a sort, away from people with whom he can converse about logic and philosophy. He is still tormented by problems in logic and the foundations of mathematics, though. Just a few days after entering the Army, he begins writing about philosophical problems again. He ponders them in relative solitude, laying out the puzzles and problems he's been thinking about and (less frequently) the progress he feels he has made, in dated notebook entries.

The moment he first glimpsed a solution was September 29, 1914. He suddenly felt that he had, as he recorded in his notebook entry that day, come upon something that contained "the solution to all my questions." He wrote:

> The general concept of the proposition carries with it a quite general concept of the co-ordination of proposition and situation: the solution to all my questions must be *extremely* simple.

> In the proposition a world is as it were put together experimentally. (As when in the law-court in Paris a motor-car accident is represented by means of dolls, etc.)

This private moment of insight was later vindicated when, still serving in the Austrian Army, Wittgenstein produced a manuscript in which he lay out his solution to the philosophical problem. He was carrying the manuscript in his rucksack when he was captured by the enemy and taken to an Italian prison camp. Both he and the manuscript survived. He sent one copy to Frege, and then he sent another copy to Russell, along with a letter remarking on how much he yearned to see it published. So, five years after the initial glimpse of the path to a solution, the publishing world was offered a German-language manuscript in which this young man, Russell's protégé, stated that he believed he had solved the problems of philosophy. Russell had made a similar declaration in *Principia Mathematica*. Wittgenstein had trouble finding a publisher, though, even with the help of prominent friends and Russell, who wrote an introduction to the book to induce publishers to take a chance on it. That introduction did spur some interest among publishers, but Wittgenstein failed to reach agreements with the few publishers who expressed interest.

Wittgenstein finally gave up trying to get his manuscript published and left it in Russell's hands. In turn, Russell, who had left Cambridge on a visit to China, asked mathematician and former student Dorothy Wrinch to try to get it published. She presented it to Cambridge University Press. They rejected it. So did many others. But when the manuscript reached the desk of the founder and editor of the *Annalan der Naturphilosophie* in Leipzig, Germany—the Nobel Prize-winning scientist Wilhelm Ostwald, whose interests had turned more and more to philosophy—it piqued Ostwald's interest. Ostwald was interested in publishing it, and he wrote back to Wrinch, asking to include the introductory essay Russell had written for Wittgenstein's *Tractatus*. In his note to Wrinch, Ostwald remarked that his interest was in large part due to his respect for Russell. However, it should be remembered that the context of that statement was a note in which Ostwald was trying to persuade Wrinch to give him Russell's introduction along with Wittgenstein's manuscript. So, although it may at first appear that Ostwald's interest in Wittgenstein was due largely to Wittgenstein's connection with Russell, the interest in Russell is probably not the whole story, considering that Ostwald turned over an entire journal issue to Wittgenstein's book-length manuscript.

One clue as to why Wittgenstein's manuscript, which had been passed over by so many, engaged Ostwald's attention is that Ostwald was then a major figure in the same German-speaking world of debates about the foundations of physics in which Wittgenstein had spent his youth. Perhaps what interested Ostwald was the kind of approach Wittgenstein had taken. Ostwald was a friend and colleague of Boltzmann's and a leading proponent of energetics. Energetics (in the sense meant by Ostwald) was a philosophical view of physics in which the concept of energy, rather than force, is made central. Boltzmann, for whom Wittgenstein had such high regard, was one of the main discussants on the subject of energetics—though Boltzmann became better known for being skeptical of the sweeping claims then being made for energetics than for publicizing its virtues. Ostwald derided the use of models as unsound and emphasized the use of equations. According to Ostwald, Maxwell's equations for electrodynamics needed no supplementation by mechanical models that (in Ostwald's view) tended to mislead; hence, the kinetic theory of gases too should eschew models of molecules in motion. Boltzmann defended the use of models in science, making the interesting observation that manipulating equations, which the proponents of energetics emphasized, was, after all, somewhat like manipulating models. Were there themes in Wittgenstein's manuscript that resonated with Ostwald's interests in philosophy and physics? The familiarity he sensed in reading it may have been due to residues of themes from an article by a physicist who had trained under him as a graduate student long before—Edgar Buckingham.

Indications that foundational works on the methodology of engineering scale models might have figured in writing the *Tractatus* come from Wittgenstein himself, who, as we have seen, emphasized the significance of thinking about the use of scale models in conceiving the *Tractatus*. He had, of course, been around scale models of various sorts most of his life. We know that in his boyhood he built and played with models of planes and other machines, and scale modeling in hydraulics was a specialty in Manchester, where he did graduate engineering work prior to leaving for Cambridge in 1911. However, we also know that Wittgenstein told friends on several occasions that it was only in 1914 that he started working out ideas about a connection between how models represent reality and how propositions do, and his writings bear that memory out. Buckingham's treatment had a peculiar flavor, putting points in terms of principles that seem more about symbols and the language used to describe things in the world than about things in the world.

If Buckingham's very general presentation of the foundation of experimental scale modeling as a method more generally applicable to statements about relations between quantities was important in the formulation of the *Tractatus*, it was more than a coincidence that the editor who wanted to publish Wittgenstein's orphaned manuscript happened to be Buckingham's doctoral thesis advisor. A suggestion such as this contains some element of speculation, just because so many things go unrecorded and so many records have been destroyed. Wittgenstein himself ordered manuscripts and notes from this period destroyed. While he acknowledged that many of the ideas in the *Tractatus* were not new, he did not bother to credit the sources of those ideas. Hence, the absence of an explicit mention of Buckingham's work doesn't indicate much either way. At any rate, Wittgenstein's manuscript first appeared in 1921 in *Annanlen der Naturphilosopie*, a German-language journal, and it was selected for publication by Wilhelm Ostwald.

Wittgenstein was not personally involved in the publication process, however, and he was displeased with the editorial changes made to the manuscript under Ostwald's editorship. The search for an English publisher for the *Tractatus* continued. When finally the manuscript was to be published in a bilingual German-English edition, Wittgenstein made sure he was closely involved in the English translation. He had trouble choosing a title for the English translation; after considering various suggestions by friends and finding fault with most of them, he settled on *Tractatus Logico-Philosophicus* for the English version of the *Logisch-Philosophische Abhandlung*.

What happened after the publication of *Tractatus Logico-Philosophicus* is another story—a story in twentieth-century philosophy and intellectual history. Wittgenstein's *Tractatus* not only went on to become carefully studied in Cambridge, but soon there were reverberations in Vienna with the establishment of the Vienna Circle, a regular meeting in Vienna that gave birth to something that had the zealous flavor of a major movement. The movement I mean, often known by the label "logical empiricism," eventually migrated along with many of the group's members and participants to the U.S. (notably, Rudolph Carnap (the University of Chicago, and then UCLA), Herbert Feigl (the University of Minnesota), Phillip Frank (Harvard), and Hans Reichenbach (UCLA)) and to England (A. J. Ayer (Oxford) and Frederick Waismann (to Cambridge, and then Oxford)). The story in this book, however, is concerned with what led up to the publication of the *Tractatus*—in particular, that crucial moment in autumn 1914 when, as a soldier in the early months of war, Wittgenstein said he had the idea for the fundamental thought of the *Tractatus*.

§§§§

The story of Wittgenstein tackling the problems of logic and philosophy and the story of the Wright Brothers tackling the problem of heavier-than-air flight have usually been told as two unconnected stories—one taking place mainly in Wittgenstein's Vienna and Russell's Cambridge, and the other in the American Midwest of the Wright Brothers and on the desolate, windblown sands of Kitty Hawk. But there were connections. One such connection was the community of physicists, which included Edgar Buckingham and Ludwig Boltzmann.

In 1894, when Wittgenstein was about five years old, Boltzmann gave a talk in Vienna, in which he urged research into heavier-than-air flight. The lecture (included as an appendix to this book) was included in the collection of his popular scientific writings that was published in 1905, just when Wittgenstein was so full of admiration for Boltzmann that he hoped to study physics with him. Boltzmann was concerned that Germany was falling behind England in research into heavier-than-air flight. Solving the age-old problem of heavier-than-air flight, Boltzmann said in that lecture, would be like Gauss's solving an age-old question in algebra, the "Kreisteilung Problem." Just as the Kreisteilung Problem had been considered insoluble because it had resisted solution for so long, so had the problem of flight been considered insoluble, because so many had tried and failed. However, the situation had changed of late, he said. It was now clear that the solution was close, and his audience was in the position Gauss was in when he realized the Kreisteilung Problem could be solved. The most important research needed, he said, could be conducted with a simple child's toy—a kite.

As it turned out, the problem of flight was solved by the time Wittgenstein was ready to conduct serious research in the field. But his first step into aeronautical research—going to England to design, build, and fly kites—did eventually bring him to the problem he felt it was his destiny to solve. He was directed to read works by, and then go to study with, Bertrand Russell at Cambridge, England, and, eventually, to answering one of the related questions in the logic of science that Boltzmann had contemplated: the relationship between models and equations, or, put another way, between models and statements. It took another step, beyond collaborating with Russell, before he came to think that a proposition was "a picture," or "a model of reality," and that understanding how would solve the problem. The next step would be his own. He would break away from the presumption almost everyone else had made that it is the form of our sentences or equations that mirrors the world, and see that instead it is our language itself that mirrors the world.

Chapter 1

Toys to Overcome Time, Distance, and Gravity

L udwig Wittgenstein was born near the city of Vienna, Austria on April 26, 1889. The household into which he was born already had four sons and three daughters; he was his parents' eighth and last child.

His immediate family provided him an uncommon vantage point, for his father was Karl Wittgenstein, the immensely wealthy industrial magnate of the European steel and rail industry. Though both he and his wife were children of successful businesspeople, Karl Wittgenstein's empire had not been inherited. He had entered industry in his youth as an engineer designing new steel mills. He had rapidly acquired responsibility by promotion, then amassed immense personal wealth by investing in coal, iron, rail, and steel concerns, even establishing new steelworks. It was a time of industrial expansion in Europe, and advances in technology were common topics in newsmagazines. Technological innovations would have been of interest in such a household. But so too were literature, art, and music—especially music. Poldy Wittgenstein, Ludwig's mother, was a pianist, and the Wittgenstein home in Vienna contained several grand pianos. At the time Ludwig was born, listening to music meant hearing live performances by musicians, often at private gatherings, and the extravagant Wittgenstein home at 16 Allegasse in Vienna was the venue of many such "musical evenings." Johannes Brahms was a frequent guest in the household, as were Clara Schumann, Richard Strauss, Gustav Mahler, and many other composers.

The year Ludwig Wittgenstein was born, nearby developments already underway portended two wondrous changes of the coming century: the advent of controlled heavier-than-air flight and the mass production of musical sound recordings. Before they brought about radical social and cultural transformations, though, these innovations appeared in Europe in the form of children's toys. Both a rubber band-powered model helicopter-like toy and a working toy gramophone with which music could be reproduced from hard discs appeared in Europe in time for Ludwig's childhood. And, as we shall see, not only were both innovations part of his childhood, but both reappear in his work as an adult.

On December 17th, in the year Ludwig Wittgenstein was born (1889), and in the city in which he lived (Vienna), Brahms recorded himself playing the piano. The recording was made on a wax cylinder in the apartment of his friend Dr. Fellinger. Though extremely fragile, the recording has been preserved. It was transferred from a wax cylinder to a gramophone disc in 1935 and is now available as an MP3 file on the Internet. You can now download it onto an iPod or a similar device and carry it around with you to play back whenever you like, re-creating the sound waves Brahms made in that apartment in Vienna the year Wittgenstein was born, before there were even gramophone discs. That such care has been taken to preserve it reflects how precious even a brief live recording was then.

Until then, the only way to communicate a musical composition other than by hearing a live performance was by sheet music (i.e., the musical score), and the publication of sheet music was a lively business. One visitor to the Wittgenstein household reported that "From time to time superb autograph manuscripts of the Viennese musical classics were to be seen lying around open as one wandered about ..." Publications of the sheet music of new compositions generated the kind of interest that new releases of musical compact discs do today.

That a sound recording capable of being played back at least once was possible in principle had been proved well before 1889, but the recordings were fragile. Both the number of recordings of a single performance that could be produced and the number of times each recording of it could be played back were, until just a few years before Ludwig's birth, very limited. In addition, the quality of the recording was less than exact reproduction; in the early technologies, the reproduced sound was distorted and allowed only recognition of what was being said, not of who was speaking.

Emile Berliner's technology of hard gramophone discs eventually beat out Thomas Edison's use of cylinder recordings in his phonograph. In fact, gramophones eventually came to be called phonographs in the U.S. Berliner eventually developed a method whereby the quality of the reproduced sound was so good he described it as an "exact reproduction," and with which an unlimited number of discs of a single performance could be produced. Berliner was a German who had emigrated to the U.S. in 1870 at the age of nineteen. In 1888, ten days after he had invented the improved gramophone (but had not yet settled on rubber discs), he demonstrated it at a meeting of the Franklin Institute, and remarked on the excitement of hearing recordings of voices of people from whom we are separated by time or distance. He closed his presentation of the improved gramophone with speculations about its practical applications: "… whole evenings will be spent at home going through a long list of interesting performances. Who will deny the beneficial influence which civilization will experience when the voices of dear relatives and friends long ago departed, the utterances of the great men and women who lived centuries before, the radiant songs of Patti, Campanini, Nieman and others … can be heard and re-heard in every well-furnished parlor?"

Although he lived in, loved, and developed his invention in America, in 1889 Berliner traveled back to Germany to present his improved gramophone to the Electro-Technical Society of Berlin, at their invitation. While in Germany, he also arranged to have some single-sided gramophone discs produced there in late 1889, but sound quality was still an issue. A German toy manufacturer showed interest in the device, however, and the next year, in July 1890, it began to market a toy gramophone cranked by hand that was capable of reproducing music from hard 12.5-centimeter discs. In addition, it produced a "talking" doll that used a smaller, 8-centimeter disc. These were also imported to England for a short time. Berliner returned to the U.S. the next year to further develop his invention and set up companies to manufacture it.

So, in 1889, the year Ludwig Wittgenstein was born, the first mass produced gramophone discs in the world were produced in nearby Germany. The next year, a working hand-cranked toy gramophone was sold in Germany. It would not be long before songs could be reproduced in infants' nurseries and the living rooms of America as well as in Europe, using a gramophone and analogue recordings on rubber discs. The advent of accurate, durable, mass-produced sound recordings of musical performances must have been especially significant in the household into which Ludwig was born, since music played such a prominent role there. The gramophone, which enabled anyone to conjure up great musical performances, would surely have been of great interest.

Brian McGuinness, the author of *Young Ludwig: Wittgenstein's Life 1889–1921* and for many years a philosopher at Oxford, meticulously researched Wittgenstein's early years, often working with members of his family. He writes of the attitude toward music in the household: "All the emphasis was on the expression of the musical idea and it was this that was discussed with a minimum of technical terms and in the vocabulary of cultivated and perceptive participants in the long Allegasse analyses that followed each Vienna Philharmonic Concert."

The invention was conceptually interesting as well as having a major practical impact, for now there was a way to represent a particular musical performance: by the grooves or lines in a rubber disc, from which sound could be reproduced by the motion of a needle moving in response to the sounds produced by the musical performance.

The example of alternative durable representations of a musical composition—a written score consisting of marks on paper and an analogue gramophone record consisting of grooves in a rubber gramophone disc—reappeared years later. When Wittgenstein had grown into a young man concerned with solving problems in logic—specifically, the question of how a picture or model can depict something else—he used the example of the lines on a gramophone record to illustrate the relationship between "language and the world." Then he reflected on the relationship between four different things: a musical thought, the musical score of a symphony, the sound waves made during a symphony performance, and a gramophone record of the symphony performance. He remarked that they "all stand to one another in the same internal relation of depicting that holds between language and the world. They are all constructed according to a common logical pattern." What he emphasized then was something much emphasized in both popular and technical accounts of the gramophone in the days when the technology was new: the processes by which one of these four different things can be produced from another. The lines on the gramophone record and the musical notation of the musical score of a

symphony are both visual, and Wittgenstein may have very early on contemplated how such different things could both be of the same musical
composition.

Much later, as a young man, Wittgenstein contemplated translation
between the gramophone lines and the musical score as a sort of translation between languages. The similarity between these very different things
was accounted for in terms of the processes by which one of them could be
derived from the other. There are four processes to think about: (i) the
process by which the musician produces the score from the symphony, (ii)
the process by which the musician produces the symphony from the score,
(iii) the process by which the lines are produced from the sound waves,
and (iv) the process by which the sound waves are produced from the lines.

The process by which the musician "reads" the musical notation of the
score and "hears" the symphony, imagines "the musical idea," or at least
understands what and how instruments are to be played to produce a symphony, is a matter of skill. Then there is the process by which the score can
be produced by a musician who hears the symphony, or hears the sound
waves produced from the gramophone record grooves. As we shall see
later, as a young adult writing on the logic of depiction, Wittgenstein
thought of this process as something that would be carried out by a musician who already knew how to read a score, who already knew how to
"obtain the symphony" from it. The musician would use the same "rule,"
he said, to derive the score from the symphony—to put a heard symphony
into the language of musical notation. He referred to this as "the law of
projection which projects the symphony into the language of musical
notation." I take it that here he used projection in its mathematical sense.
For example, if we are interested in how much floor space an item will take
up, we only ask what the projection of its dimensions onto the plane of the
floor is; it is not necessary to mention details about its shape in the vertical
direction. Similarly, for musical notation, there may be features that a particular symphony has that are peculiar to that performance, and are not
part of the musical composition; these would not be captured by, or projected into, the musical notation. It is notable that Wittgenstein thought of
the ability to read the score as the *primary* human skill, and the recording
of a symphony in musical notation as something done in virtue of possess-
ing that skill.

Then there is the process by which the gramophone record is produced, whereby sound waves evoking the original symphony performance are produced from the lines on the gramophone record, and the inverse process; i.e., the process whereby the gramophone lines on the gramophone disk are produced. This latter process is relative to human capabilities as well, for, though the recording devices involved only mechanical processes, the first recording devices were modeled on the human ear. It makes sense that, for a sound recording meant to capture a musical performance to be listened to by humans, it is only important to capture what humans can hear. Here it is more evident that what is recorded is a "projection" of the symphony into the language of the gramophone lines. Ultrasonic frequencies outside the range of human hearing that are produced in a symphony production are not relevant to the gramophone record of the performance. The process whereby gramophone records are produced, however, does *not* figure in the account of translation Wittgenstein gives in the *Tractatus*, written when he was in his twenties. Close attention to the text in which he describes what provides the means of translation reveals the absence of any mention of the mechanical process of recording sound waves:

> 4.0141 There is a general rule by means of which the musician can obtain the symphony from the score, and which makes it possible to derive the symphony from the groove on the gramophone record, and, using the first rule, to derive the score again. That is what constitutes the inner similarity between these things which seem to be constructed in such entirely different ways. And that rule is the law of projection which projects the symphony into the language of musical notation. It is the rule for translating this language into the language of gramophone records.

Instead of the process of recording lines or grooves in the gramophone record, what figures in Wittgenstein's account of translatability between the musical score and the gramophone record here is instead the process whereby sound waves are produced from those lines, or grooves. (And, after that, the human skill of putting the heard symphony into musical notation.) This is striking, for, actually, the visual record of sound as wavy lines predated the production of gramophone records meant to be used to reproduce sound waves.

That other, earlier, visual record of sound is yet another kind of representation associated with the invention of the gramophone that was actually the springing-off point for the development of the gramophone record. It was called a phonautograph, and Berliner begins accounts of his own invention, the gramophone, by describing it. Phonautographs produced by a machine also called a phonautograph seem to have been well-known at the time, for Berliner speaks of "Scott's phonautograph" as if assuming audience familiarity with it, and another paper on the principles of the gramophone by a Professor Houston refers to it as "the well-known phonautograph of Leon Scott."

Scott's story was poignant: his family was too poor to give him an advanced education, and he was apprenticed to a printer. His work involved printing the transactions of scientific societies, which he read in the course of copyediting. He got to know some of the scientists whose work he printed, and he began corresponding with them about their work. These pursuits led to Scott's inventing a machine that would produce a visual record of sound. Scott's illustration of his invention shows a person performing on a musical instrument in front of the machine, and the machine, built on the model of the human eardrum, producing a series of wavy lines distinctive of the performance. The record consisted of lines caused by the motion of a membrane, which was in turn caused by the sound waves produced by the musical instrument. The sound records were white wavy lines scratched in a blackened surface formed by a smoky film on paper. The point was that they were distinctive marks corresponding to sound waves, and that, like any other two-dimensional icon, they could be reproduced without limit by a printing process.

The point was to have a method of recording sound, somewhat like present-day seismographs record waves traveling along the Earth's surface. Some put special significance on the production of wavy lines that were geometrically similar to the sound waves that produced them. Berliner remarked that the hard zinc disc made in his process "becomes a picture of sound waves which, though slumbering in a bed of hard metal, is ready at any time, even centuries hence, to burst forth into the soft cadenzas of word and song, the ripple of laughter, the strains of martial music, as well as the melancholy and imploring drag of the organ-grinder's tuneful melody." However, geometric similarity to the actual sound waves was not essential to the goal of producing some sort of graphical or iconic representation of sound. Hence, the role of the lines as pictures of sound in virtue of their similar visual appearance and pictures of sound in virtue of the ability to produce sound waves from them diverges for Berliner. (As we have seen, Wittgenstein was sensitive to this point, too, for his account of similarity between gramophone lines and other representations of sound does not appeal to geometric similarity.) In March 1857, Scott was granted a patent for "a method of drawing or writing by sound, and for multiplying the result of this graphically with a view to industrial applications." The same kind of device was also called a logograph. Scott did not attempt to use the graphical representations to actually produce sound, but Edison and then Berliner subsequently saw the potential of such a complementary process. This complementary process, which Scott seems not to have noticed or cared about, led to the development of Edison's phonograph and Berliner's gramophone.

Thus, for a while, there were phonautographs, or visual records, of sound, and these were well known before and during Wittgenstein's childhood. Berliner remarks that Scott's phonautograph "is described in every book on physical science," and, in fact, Berliner talks about using printed phonautographs as a means of conveying the gramophone sound recording. The ability to produce sound from them, while still regarding them as two-dimensional visual representations, is illustrated in a particularly colorful way in his fanciful speculation that "We may then have dinner-sets, the dessert-plates of which have gramophone records pressed in them, and which furnish the after-dinner entertainment when the repast is over." This is immediately followed by the speculation that "Gramophone plaques with the voices of eminent people will adorn our parlors and libraries."

Likewise, the dual aspect of a gramophone disc—both like the written word yet also able to produce the sound represented by the written word— was a novelty not lost on its inventor. Berliner wrote "I am carrying on a vocal correspondence with my friends in Europe, by means of small gramophone discs, which can be mailed in a good-sized letter envelope.... I could cite a number of instances where persons have been made happy by hearing and recognizing the voices of loved ones whom they had not seen in years, and the owners of which were thousands of miles away."

It is notable that in his discussion as an adult about the gramophone record, Wittgenstein does not include the kind of graphical record that a phonautograph is among the group of things that have the same "logical form" as the musical score—even though Scott was aiming precisely at providing a graphical representation of sounds. It makes sense that Wittgenstein does not include Scott's phonautograph, however, given his explanation there of what logical form consists in, since there was no way to produce any of the other records of sound from a phonautograph unless there were some kind of playback mechanism. Of course, by the time gramophones were in existence, he would have been aware that it was always theoretically possible to develop a machine to play back a phonautograph record, but the existence of a playback mechanism would essentially make the phonautograph record a gramophone record—which is what he does use to illustrate his points about logical structure and pictorial representation. But it is also striking that the crucial aspect that Wittgenstein cites as accounting for logical form in the philosophical treatise he writes as a young man—"there is a rule by which one could reconstruct the symphony from the line on a gramophone record"—is precisely the advance in sound recordings that was exhibited in the toy gramophone that premiered in nearby Germany just after his birth.

There is another point important in Wittgenstein's philosophical writings on depiction that is consistent with making a distinction between Scott's phonautographs seen as lines on paper and Berliner's view of exactly the same object as something from which sound waves can be produced. Scott's phonautographs are not pictures of sound for Wittgenstein, since they do not include the concept of a process whereby the sound can be produced from the lines; Berliner's notion of the lines, which includes the notion of a process whereby sound waves can be produced from the lines, does. Wittgenstein not only made the distinction for pictures of sound, but he saw in it a point about all pictures and thus about the very nature of depiction: after explaining that a picture is a fact, he writes that "a picture, conceived in this way, also includes the pictorial relationship, which makes it into a picture."

A story about a fraud perpetrated on a count in the Vienna Woods in the summer of 1888 highlights the role of a playback mechanism. In his account of the year preceding Wittgenstein's birth in *A Nervous Splendor: Vienna 1888/1889*, Frederic Morton writes of an unexpected visit paid to a count whom he describes as "the principal performer in the amateur musicals given in his house":

> During the hot months of 1888 … a coach rolled into the leafy driveway of a villa in the Vienna Woods. It belonged to Count Walter H…. A gentleman stepped out, excellently cravated, and handed a footman his calling card. It said *Philip H. Elkins, Esquire, New Jersey, U.S.A.*

> Admitted to the Count's presence, Mister Elkins introduced himself as the chief European representative of Thomas A. Edison Enterprises of New Jersey [and said that] Mister Thomas Edison was planning a phonographic gallery of famous great voices of the nineteenth century. At Mister Edison's request he had therefore brought along an Edison machine in the hope that the Count might be kind enough to let the machine record the art coming from the Count's throat.

> The Count was most cooperative. With the help of two of his footmen, a heavy American-looking machine, bristling with tubes and wires, was dragged out of the coach, over precious carpets, into the music room. Here the Count sang feelingly his favorite aria, "Se vuol ballare," from *The Marriage of Figaro*, while Mister Elkins kept adjusting levers to accommodate the remarkable volume of the Count's voice.

When the Count asked to hear his voice played back, however, he was told that the only such machine in existence was in New Jersey with Mr. Edison. The Count suggested building a "playback apparatus" in Vienna, giving him "two hundred and fifty florins to get the work started, plus a fifty-florin licensing fee to Mister Edison." Whereupon

> Mister Elkins then carefully guided the footmen as they heaved the wires and tubes out of the house and into the coach, climbed into the coach himself, waved his hat, and was never heard of again.

Such a story naturally raises questions about when a sound recording really should be counted as a sound recording. When is it a fraud? When is it, though intended to be a recording, useless or meaningless? When is it correct to say that it is in some sense a record of a performance of an aria? And what does being a recording consist in?

Given the times, the city, and the household in which he was born, the gramophone was undoubtedly important to Wittgenstein in his childhood, and we know that it was important to him as an adult, too, for it is reported that Wittgenstein "when listening to music on the gramophone put the needle back repeatedly to some musical transition from which he wanted to extract everything." The question of depiction could well have arisen much earlier in his life, in the form of the difference between a gramophone record and a phonautograph. But puzzles about the relation of these pictures of sound to the musical notation used in a musical score might not have been so explicit. Perhaps all that he got out of this from his childhood years, besides the preceding point about pictures, was one example of alternate depictions of a musical composition; even so, his observations about it resonated decades later in working out problems in logic and philosophy.

Another toy common in Europe during Wittgenstein's childhood was one that actually flew: an elastic-powered, helicopter-like toy. George Cayley had developed the toy in England in 1796 as a small model, drawing on even earlier versions. Cayley was an independently wealthy Englishman who somehow became convinced that heavier-than-air flight was possible, and he is credited with the very concept of an aircraft with fixed wings. Because his work on airflow over inclined wings was the first accurate research done on the subject, he is often referred to as "the father of aeronautics." Cayley's model helicopter design was later perfected and made popular in France by Alphonse Pénaud. The toy was very popular in Europe during Ludwig's childhood, and he almost certainly would have been familiar with it. The Wright Brothers described the toy they had played with and tried for years to make larger copies of as a European toy called a "bat." Orville Wright was adamant in his recollections that it was the toy based on Pénaud's design (and not another one known as a "butterfly" and often confused with it). There had been toy helicopter designs for centuries, but Pénaud's version was rightfully well known and much reproduced; its performance remained unsurpassed. It has been credited with inspiring a whole generation of children to become interested in flight. Octave Chanute described it in his 1894 work, *Progress in Flying Machines*, as "the best of its kind," remarking:

Pénaud's flying screw, which is called by the French a "Helicoptere," consists of two superposed screws rotating in opposite directions, and actuated by the force of twisted rubber strings.... These models, when built in varying proportions, would either rise like a dart to a height of some 50 ft., and then fall down, or sail obliquely in great circles, or, after rising some 20 or 25 feet, hover in the same spot for 15 or 20 seconds, and sometimes as many as 26 seconds, which was a much longer flight than had ever before been obtained with screws.

Chanute mentions that the models behave differently depending on "varying proportions." If by "varying proportions" he meant models of different sizes (the proportion being the ratio of lengths in the original version and another one made of a different size), then he is simply saying that the device behaved differently depending on its size. That would explain why playing with different-sized models of the toy, which the Wrights said occupied them for years, might be so worthy of study. The toy used stored energy (the energy in the rubber band), so that in a way it was self-propelled, or, as one historian put it, it had "a perfect lightweight powerplant." It was inherently stable due to its clever design, the rotating propellers rotating in opposite directions. Chanute reports that, with a suitably light engine, a size large enough to carry humans might have actually flown appreciable distances. So the toy held the promise of humans being able to defy gravity, and it illustrated the puzzling effect of size.

There was more reason than the behavior of Pénaud's "bat" toy, exhilarating as it was, for his contemporaries to feel that gravity-defying human flight in heavier-than-air machines was possible, to feel that it really was so close that it might possibly be achieved. Pénaud was a serious aeronautical researcher whose life was cut short at age 30 by his suicide after a series of disappointments, the last being finding out that the expected funding for his next experimental flight would not be forthcoming after all. Chanute summed up how things stood by the end of his life: "M. Pénaud was criticized, decried, misrepresented, and all sorts of obstacles arose to prevent the testing of his project. He lost courage and hope, his health gave way, and he died in October, 1880, before he had reached 30 years of age." Pénaud had been born with a degenerative hip disease, a condition that was both disabling and painful. He had to use crutches to get around, but he designed flying machines that soared on their own. He left behind a promising aircraft design called a "planophore," for which there was good reason at the time to believe it might have flown. The helicopter toy was special because it worked so well, but the planophore was more distinctively his. It was special in that, even as a model, it established a landmark in aviation history. Richard Hallion describes the event, which took place in the presence of a large audience, and so had become famous by the time Wittgenstein was born:

> ... the young Frenchman ... had his sights set on nothing less than developing a full-size airplane, and achieving that goal would require a number of technological demonstrations. To Pénaud a practical airplane would have to incorporate a high degree of inherent stability—the ability to fly in such fashion that a pilot did not need to manipulate the controls constantly to keep it on a steady course. Experimentation had led him to develop a configuration that he believed would work, and now he was ready to demonstrate it to the public. [...]

> That morning in Tuileries, his compatriots watched, intrigued, as Pénaud slowly turned the propeller through 240 revolutions, winding the rubber cord tighter and tighter. Then he held the model at head height and let go of the propeller, and as it immediately began spinning with a slight buzzing sound, he launched the model horizontally in the air. As he wrote later, "For an instant it started to drop, but then, as its speed picked up, it flew straight away and described a regular movement, maintaining a height of 7 or 8 feet, covering a course of 40 meters [approximately 131 feet] in 11 seconds." It had followed a slightly curving path, flying several gentle circles from the propeller's torque until the rubber bands fully unwound, and, its power exhausted, it smoothly glided to earth. Stunned, the onlookers quickly measured the distance. The first significant powered flight of a heavier-than-air flying machine was history, and young Pénaud was the talk of the aeronautical world.

Hallion remarks on its significance: "what [Pénaud] achieved that August day in 1871 was no less than the answer to the question 'Can an airplane fly?', a question that dated to the very dawn of interest in mechanical flight." The historical significance was that it changed the state of the art, in that the open questions had shifted from one concern to another:

Critics could no longer doubt that an airplane *could* fly; rather, the issue would be one of scale, involving two critical questions: *Can an airplane be built with an engine of sufficient power to lift a human aloft?* and *Can the operator control it?*

What Hallion says here about silencing the critics may have been true in France, but there were certainly critics in other countries, such as the U.S. and Britain, who did doubt that an airplane could fly, and with every new failure they were more assured of the validity of their doubts. However, in Europe, children had Pénaud's "bat" toy to play with in their own hands and the story of the very public success he had had with the planophore in their heads. When Pénaud committed suicide, he was dramatic about it: he put drawings of all his inventions in a coffin, which he arranged to be sent to the would-be benefactor whom he felt had let him down.

Many people outside France were inspired by his success, in spite of the substantial prejudice against practical heavier-than-air flight that still existed. Inventors such as Emile Berliner (who had invented and manufactured the gramophone), Alexander Graham Bell, Hiram Maxim (inventor of the machine gun), and, of course, the proprietors of the Wright Cycle Company devoted much of their time and parts of their fortunes to building a practical flying machine after their success at other pursuits had provided them with financial means sufficient to indulge their passion for flight. Samuel F. Cody's stage pursuits—massive efforts involving his whole family and portraying himself somewhat fraudulently as connected with "Buffalo Bill" of the American West—became largely a means of financially supporting his interest in flight, particularly the development of his man-carrying kite. The scientist Samuel P. Langley and the distinguished engineer Octave Chanute already had established careers in their own fields when, late in life, they devoted their energies to promoting flight research and building their own experimental aircraft. Lawrence Hargrave (in Australia) used his own fortune to invent an inherently stable kite and an engine to power it, and Alberto Santos-Dumont (a South American living in Paris) drew from his inheritance to develop a heavier-than-air machine based on Hargrave's designs to compete in French competitions. Although there was sometimes the hope that success would pay off, either in patents or in prizes at competitions, money was seldom if ever the motivation for participating. When asked, the entrants almost always explained that they were pursuing flight because of an irresistible desire to know, to discover, to invent what was waiting to be known, discovered, or invented. However, they did not all agree on which of the two remaining questions about flight—power or control—deserved attention most.

The question of the effect of a machine's size on its performance can be asked about any number of toys that are miniatures of larger craft or machines: a boat, a building, a simple bow and arrow. But, in Wittgenstein's youth, it was especially striking for aircraft, for there were models that worked well on a small scale but failed or behaved qualitatively differently on larger scales. There was the European "bat" toy, as I have said, which, when imported to America a few years before Wittgenstein's birth, made flight research irresistible to the Wright Brothers, around seven and eleven years old at the time. Besides the toy models, there was Pénaud's spectacular model planophore of a design meant to be built on a scale that could carry humans, and for which the question of the performance of a full-size version was still open.

Though not immediately obvious, the questions raised by Pénaud's pursuits, of which the ubiquitous "bat" toy was a constant reminder, are analogous to the questions about representation and language occasioned by the gramophone. Just as the musical score and the lines on the gramophone represented the symphony performance in some way, so the model of a flying machine was supposed to, in some way, represent the performance of a full-size one. In the case of aircraft, the skill required to "project" the features of a full-size model onto a smaller one of some sort was often claimed, even though the reverse of that process, by which an experienced model builder would be able to produce a practical full-size flying machine from a model, was often revealed to be faulty.

We can see that the pressing question was about this process and its inverse: the experimenter's task of building a model whose behavior would reflect (as most models did *not*) the behavior of an imagined full-size flying machine, and the inventor's task of building a full-size version of a model in such a way that it mimicked the model's performance. These seem to be human activities involving skill, as are the musician's activities of reading and writing a musical score. Yet, it also seems that, once built, the connection between an experimental mechanical model and what it models ought to be as mechanically determined as that between a gramophone record and the sound waves it produces. The analogy does not of itself provide an answer to how the representation is effected, does not say more about the logic of depiction than that models somehow depict. What an analogy between how models depict and how gramophone records depict does do is give a kind of representation that is an alternative to language. It provided Wittgenstein with something to reverse, as in von Wright's account of the key insight Wittgenstein had in late 1914, mentioned in the Preface. He could ask: rather than assuming that a model works on analogy to a statement of a language, why not think about whether language works on analogy to models—whatever that account may turn out to be? This would put him on the alert for a satisfying account of experimental scale modeling.

As numerous works of intellectual history have pointed out—most notably, Toulmin and Janik's *Wittgenstein's Vienna*—philosophy of language and revolutionary approaches to symbolic representation in music and art figured prominently in the Vienna in which Wittgenstein was born and spent much of his youth. Toulmin and Janik cite Schoenberg's twelve-tone system of music in particular, comparing Schoenberg's "breaking through the limits of a bygone aesthetic" (as he himself put it) to the work of the modern logicians De Morgan and Boole, which they see as analogously "breaking through the limits of a bygone logic." They even see Schoenberg's *Harmonielehre* as closely analogous to Whitehead and Russell's *Principia Mathematica*, since both, they say, are "compendious expositions of a new logic." But it is also true that, in his own ruminations about a theory of symbolism, the grown Wittgenstein eventually brought to the solution consideration of things not occurring frequently in those intellectual discussions—the lines on a gramophone record and experimental scale models.

The toy gramophone and aeroplane portended important changes in the lives of adults and therefore were not regarded condescendingly, as toys such as dolls or wooden trains might be. While still available commercially only as toys, they were discussed in lectures about the science underlying the developing technology, as in Berliner's talks to engineering societies about how energy considerations in reproduction of sound had led him to the new rubber disc technology, and in Lilienthal's and Chanute's discussions of the forces involved in the behavior of various toy helicopters, aeroplanes, and gliders. These playthings from Wittgenstein's childhood were astounding to people of all ages, in terms of the way in which they employed basic science in challenging the presumptions of everyday experience. One toy illustrated how the time and distance normally separating a distant or absent hearer from a speaker's voice could be overcome with a machine that re-created it from a pattern of lines. The other gave hope to an equally romantic notion: the ability of a vehicle that was heavier than air to defy gravity and fly in the air, to carry a person when and to where one wished, at will. Like many others of his generation, Wittgenstein wanted to design, build, and fly his own airplane; like a few of them, the desire shaped his career choices as a youth. But there was more to these childhood experiences than the directions in which they led him as a young man: the reflections they occasioned and the examples they provided were resources upon which Wittgenstein could, and did, draw as a philosopher.

Chapter 2

To Fly Like a Bird, Not Float Like a Cloud

T hough he was born just a bit too late to have a chance at solving the problem of practical human flight himself, the boundaries of Ludwig Wittgenstein's childhood and youth framed a window of time well suited to observe the saga of its solution. Instead of a saga told in terms of one generation of a family carrying on what the previous had begun, the saga of the invention of flight unfolded in terms of one flight pioneer carrying on the work of a predecessor—more often than not, one who had met with an untimely death. Even before Wittgenstein's birth there had been such a predecessor: Alphonse Pénaud, the pioneer in powered heavier-than-air flight whose publicly observed experimental model of aeroplane flight had convinced many of the practicality of heavier-than-air flight, and who had invented the toy helicopter that was so popular in Europe during Wittgenstein's childhood, had, as we have seen, committed suicide years before.

In 1889, the year Wittgenstein was born, Otto Lilienthal's book *Der Vogelflug als Grundlage der Fliegekunst* (*Birdflight as the Basis of Aviation*) was first published. It marked the end of Lilienthal's preliminary research into flight and announced his plans for the next step: experimentation with a human-carrying winged apparatus. When Wittgenstein was seven years old, Lilienthal died from injuries received when one of those gliding experiments ended in a crash. The event was widely reported. Lilienthal life's work was recounted in the worldwide press following the dramatic circumstances of his death, and it would have been a vivid and memorable event in Wittgenstein's childhood when he was at an impressionable age.

Across the Atlantic Ocean, news stories on Lilienthal rekindled interest in his work in the minds of two brothers, Wilbur and Orville Wright.

During Wittgenstein's first year at the Realschule (a kind of high school meant for students who planned practical careers and so did not prepare them for entrance to university), those two American brothers working closely together to solve the problem of human flight. Conflicting rumors of what they had accomplished circulated in Europe in the years that followed, as the Wright Brothers kept their invention under wraps while waiting for their patent application to be approved. They were doubted in large part because they would not participate in exhibitions; many thought that if they had really achieved what they said they had, they would be eager to show off their invention in competitions. However, the Wright Brothers considered exhibiting and competing for prize money rather vulgar. They were more concerned about getting a patent before convincing the world at large that they had built and tested a practical flying machine.

By the time Wittgenstein had completed his studies at the Realschule at about sixteen years of age, rumors of the Wright Brothers' success, though still widely and openly doubted in Europe, had gained credibility among people who followed aeronautical research seriously. Finally, in the summer of 1908, just a few months after Wittgenstein left the Continent to study aeronautical engineering in northern England, the Wright Flyer was brought to Europe, and Wilbur Wright flew it spectacularly well in exhibition after exhibition to sold-out audiences. It was clearly not simply an acrobatic trick—nor were the photographs that had preceded the arrival of the real thing in Europe illusions. In that year, the whole world came to see that the problem of practical heavier-than-air human flight had indeed been solved.

All of this development—from the publication of Lilienthal's book advocating a research plan to its successful completion by the Wright Brothers—took place during the years between Wittgenstein's birth and the summer he finally got a chance to begin aeronautical research at the age of nineteen. His letters home indicate that he thoroughly enjoyed the work of designing, building, and flying kites that summer. The photograph shown on the cover of this book is of Wittgenstein and his engineering colleague William Eccles with a kite, probably designed and built by them.

The kite shown is a box-style kite, a style first developed in the 1890s by an Australian, Lawrence Hargrave, as mentioned earlier. Samuel F. Cody later developed it for practical scientific and military uses; the shape of the kite in the photograph shows some influence of Cody's unique "extended wing" design. The kites Wittgenstein designed and worked with were used for meteorological observation. The kite flying was research associated with the program in aeronautical engineering he was to begin in England that fall, the autumn of 1908—just at the point in time that it became settled to everyone's satisfaction that the problem of human flight had been solved.

Besides the exhibitions in continental Europe in the summer of 1908 at which attendees were reported to be "startled" by Wilbur Wright's sustained, agile flights, there was another event that took place in 1908 much nearer to where Wittgenstein was then studying aeronautics. The first heavier-than-air powered manned flight in Britain took place the fall Wittgenstein began study as a research student in aeronautical engineering in Manchester. In October of 1908, Cody, already famous for his success in inventing and building "man-lifting" box-style kites for the Army for surveillance purposes, built what he called a "powered kite"—that is, an aeroplane—and successfully flew it for a quarter-mile.

Samuel F. Cody was born in America but moved to England in 1889, the year Wittgenstein was born. At the time, he earned his living by putting on a Wild West Show, and he brought the show to England. He shortly met the love of his life, Lela, fifteen years his senior and married with several sons. They became inseparable, and Cody integrated Lela and two of her sons into his show with dramatic stunts that pushed the limits of what audiences were used to seeing on stage. All three newcomers to the show enthusiastically took up sharpshooting practice. Garry Jenkins' scholarly but spicy biography of Cody reports that one stepson "learned to fire accurate shots while hanging from a trapeze," and the other "had been coached to fire at a target while Cody held him upside down by his ankles and swung him 'like a clock pendulum.'"

Lela seems to have been naturally gifted at the sport herself. Cody's shows featured explosions, bullets fired to knock cigarettes out of ladies' mouths, horses as well as boys jumping from what looked to be dangerous heights, a horse and rider plunging through what appeared to be glass doors, and so on. In another show he wrote, it is reported that "[p]hrases like 'snakes alive' and words like 'tarnation' and 'geossifax' abounded."

But his Wild West Show, though extremely popular and lucrative, increasingly became for Cody a means of earning money to support a more important vocation: flight research. Although he was especially interested in the box kites of Lawrence Hargrave, Cody "always claimed to have gained his first experience of elementary kiting back in America, where a Chinese cook had taught him how to build and fly kites … The inspiration for building bigger and more stable kites had come 'from the big turkey buzzards I used to see in Texas … I was always struck by the way they used to float in the air, sometimes for ten minutes, without stirring a wing.' Jenkins writes:

> Experimenters like Hargrave, Wise and Baden-Powell had been thwarted by the instability of their kites and their inability to find a system by which they could control their ascents with them. Given his lack of formal education, Cody could do little to test the scientific theory behind their failures. Instead he relied on a combination of instinctive inventive genius, courage and sheer dogged determination to eliminate each problem as it arose.

Building excellent kites, like Cody's other achievements, was a family endeavor. Jenkins tells us that

> Cody's stepson Vivian later recalled … how Cody was constantly trying out new adjustments to his kites. He would … [barge] into their bedrooms before dawn. "He'd rouse us up. 'I've got a new idea. Something we must try out. Now!'" The boys would follow him to the theatre, where they would unpack the sewing machines they were by now experts at operating, then cut out the new designs Cody had come up with … "This sort of thing went on for months and months, trial and error, test, fail, try again."

The precursors to Cody's aeroplane design, his man-lifting kites, were ingenious constructions of a series of connected kites. His ingenuity was often stimulated by his intense competitive spirit. According to Jenkins' biography of him, it was like Cody to feel confident he could best everyone else who had preceded him in using kites to lift people up into the air if he brought his own skills to bear on the problem. As Jenkins put it: "With characteristic chutzpah, Cody had set out to outdo them all [Hargrave, Maillot, Wise, Baden-Powell, Eddy, Rotch, Chanute]." Cody designed kites that provided a strong lift force, all right, but, ever the mechanically gifted showman, he applied the kind of cable mechanism used in cable-car systems in mountainous regions of Europe to push the feat dramatically beyond the limits others had reached. The kites were arranged in a series and connected by cabling, the cable anchored at ground level with a heavy winch. The first and highest kite was the lead, or "pilot kite"; below that were several "lifter kites" designed to provide substantial lift. Finally, below all of these was a "carrier kite" to which was attached a sort of basket in which a passenger could ride. In 1901 Cody patented it, and he wrote to the War Office, "offering them first option on 'SF Cody's Aroplaine [sic] or War-Kite: A boy's toy turned into an instrument of war.'"

Britain was already using balloons in military maneuvers by 1880, so its Army had a section devoted to aerial operations. By 1902, Cody held the post of Chief Kite Instructor to the British Army Balloon Factory at Farnborough, England. The Balloon section of the military seems to have included kites as well as balloons. Cody did get the Army's attention, though he almost lost it at one point in the negotiations by setting a price they thought exorbitant. He later designed lighter-than-air ships for them, as well as his "Man-Lifting War Kites" intended for aerial observation. Lela, too, was involved in these aerial investigations; in 1902 she went aloft in the Cody family's invention herself.

Cody, like Lilienthal before him, would die in a crash. It happened less than five years after the historic 1908 aeroplane flight, at Laffan's Plain near Farnborough. Despite his spectacular technical successes and his lucky escapes from daring and dangerous situations, all was not glorious for Cody. Although he maintained popularity among much of the public, his career and financial situation had dramatic ups and downs. For years, the War Department gave him less respect and remuneration than he felt was his due, and he had to endure open resentment at times that Army funds were spent on someone who was not an Englishman, even after (as he reminded them) he had given up his lucrative stage shows in order to carry out the Army work for far less personal income. The flight research he carried out on his own time exhausted his funds and brought complaints about how he was spending his time rather than additional contracts for his additional inventions. He was actually forbidden by the Army to make the historic flight in 1908 that ended up being a source of pride for England, and he was later relieved of his position. Cody later become a naturalized citizen in order to be eligible to compete in aerial competitions reserved for British citizens. At the time of his fatal accident, his fortunes had turned for the better. He had won major cash prizes from the Army that had earlier given him his walking papers, in addition to still being lauded as the first to achieve powered human-carrying heavier-than-air flight on English soil. He was so popular by then that over 50,000 people attended his funeral.

The significance of the 1908 flight for which Cody was so famous and beloved lay in its being a first for England—in showing that England was at least following along with the rest of Europe and had aeroplanes and pilots of its own—and not in any crucial technical contribution Cody's aeroplane had made to the solution of heavier-than-air human flight. His aeroplane design had benefited from the information that Colonel John Edward Capper had brought back from his visit with the Wrights in 1904. Cody certainly felt that his airplane design contained improvements of his own that were significant, but, even if this is true, it was not fully appreciated or the only reason for his fame. His work on man-carrying kites was recognized as a distinctive technical achievement, and his work on the *Nulli Secundus*, the military's semirigid dirigible, was considered a more significant contribution to the nation's military forces. His most striking technical achievement, however, occurred in the summer of 1913, less than a month before his death. It filled his longtime supporters with pride and even managed to convert many of his detractors. His was the only British plane that had managed to complete all the trials set in the Military Trials carried out that summer, and he won in both the international and British-only divisions. Cody had competed under incredible odds, for his plane had crashed just three weeks before the trials, and he had to design and build another in that short time. What was striking, and in keeping with the legend of his practical genius, was that he was competing against heavily funded and scientifically sophisticated design teams. Jenkins reports that "The result provoked a mixture of delight and disgust. Many could not believe Cody's cumbersome Cathedral had outperformed the products of hundreds of thousands of pounds' worth of research and development." As we shall see, the tendency to see accomplishments in aeronautics as vindicating one or the other side of a competition between science and practical skill runs throughout the history of heavier-than-air flight and reflects a divide in the more general discipline of the study of motion in fluids.

The story of the solution to the problem of heavier-than-air human flight often begins with Lilienthal. This is not just because the Wright Brothers said it was his work that inspired them to begin their research, but because, in the book he published the year Wittgenstein was born, Lilienthal articulated the reasons for a decisive shift away from lighter-than-air aircraft to heavier-than-air aircraft. Lilienthal outlined and then carried out a rational step-by-step experimental research program. He patiently and methodically performed preliminary experimental studies on models of wings instead of prematurely rushing to build a human-carrying machine. His book provides a snapshot of his research program at the point where he is about to embark on experiments with human-carrying flying machines. The first step in experimenting with a human-carrying apparatus was the series of gliding experiments that resulted in his death.

The advantages and superior potential of heavier-than-air aircraft had been dramatized in Jules Verne's novel *Robur the Conqueror* (*Robur-le-Conquérant*) a few years before Lilienthal's book appeared. Verne's novel was published in France in 1886 and subsequently appeared in a variety of languages, including German, under a variety of titles (e.g., *Robur der Flieger* in Berlin; *Clipper of the Clouds* in London and Boston, among many others). The novel was popular among boys during Wittgenstein's childhood. In it, Verne lampooned a group dedicated to the progress of lighter-than-air flight. The group repeatedly stalled in their decision-making, unable to go in one direction or the other. The action in the novel begins when Robur shows up uninvited at a meeting in which the society's members are discussing details of the design of a proposed powered balloon for air travel. The proceedings are stalled by indecision as to whether to put the engine ahead of or behind the balloon. Robur's speech dismisses the whole enterprise of lighter-than-air flight:

> After a century of experiments that have led to nothing, and trials giving no results, there still exist ill-balanced minds who believe in guiding balloons. They imagine that a motor of some sort, electric or otherwise, might be applied to their pretentious skin bags which are at the mercy of every current in the atmosphere. They persuade themselves that they can be masters of an aerostat as they can be masters of a ship on the surface of the sea. Because a few inventors in calm or nearly calm weather have succeeded in working an angle with the wind, or even beating to windward in a gentle breeze, they think that the steering of aerial apparatus lighter than the air is a practical matter.

Robur goes on to argue against the feasibility of lighter-than-air flight ever becoming a practical means of aerial transport. He appeals to bird flight and bat flight as evidence of the superiority of heavier-than-air flight, mocking the hopes held for balloons: "A balloon, when on such a system nature has never constructed anything flying."

The narrator of Verne's novel mentions that Robur's points are the same ones made by a long list of investigators who preceded him, including Pénaud. Robur predicts: "As [man] has become master of the seas with the ship, by the oar, the sail, the wheel and the screw, so shall he become master of atmospherical space by apparatus heavier than the air—for it must be heavier to be stronger than the air!" Verne's narrator gives an overview of basic types of aircraft and a little lesson on the scientific principles involved, including these remarks:

> Nothing, in fact, is better established, by experiment and calculation, than that the air is highly resistant. A circumference of only a yard in diameter in the shape of a parachute can not only impede descent in air, but can render it isochronous. That is a fact.

> It is equally well known that when the speed is great the work of the weight varies in almost inverse ratio to the square of the speed, and therefore becomes almost insignificant.

It is also known that as the weight of a flying animal increases, the less is the proportional increase in the surface beaten by the wings in order to sustain it, although the motion of the wings becomes slower.

A flying machine must therefore be constructed to take advantage of these natural laws, to imitate the bird, "that admirable type of aerial locomotion," according to Dr. Marcy, of the Institute of France.

This view of the appropriateness of imitating the bird reflects Lilienthal's approach to some extent. His guiding inspiration that heavier-than-air flight was not only possible but preferable came from Nature— not in the sense that nothing in nature could be improved upon, but in the sense that the existence of large flying birds showed that human heavier-than-air flight was possible. Lilienthal's first known lecture on aviation was "Theorie des Vogelflugs" ("Theory of Bird Flight") in 1873, which is reported to have dealt "especially with the criticism of the balloon and with the necessity of studying bird flight." The idea of modeling human flight on bird flight was certainly not new; there had been attempts since antiquity to develop wings and attach them to humans in hopes of enabling them to fly. In fact, the idea had fallen into disrepute because there had been so many failed attempts by humans to mimic the flight of birds. Lilienthal's observations of nature were more subtle than most of those predecessors, though. He explained why it is the soaring of large birds, rather than the hovering of small insects, that artificial flight ought to emulate. He noted differences and similarities between kite shapes and naturally occurring shapes, including the shapes of pods and leaves responsible for dispersing the seeds of various plants. He noticed—and sketched in the book—the shape of linens hanging in a breeze, and of the difference in a kite's shape if the restraining crossbar is removed. He organized all these observations into a unified conclusion that gave practical guidance in building gliding machines.

Otto Lilienthal was one-half of a team of two brothers. It seems, historically speaking, a common occurrence for aviation pioneers to come in pairs of brothers working together to design, build, and fly aircraft. Modern air travel began that way, with the eighteenth-century French aeronautical pioneers Étienne Montgolfier and his brother first "sending up small paper balloons filled with heated air" and then together making a hot-air balloon large enough to carry a human; such a balloon became known as a Montgolfier.

Though history books tend to neglect them, sisters were often involved in these efforts—and mothers, too—who acted as colleagues and supporters. As Wilbur and Orville Wright's mother would do later, Gustav and Otto Lilienthal's mother provided technical advice to her sons and encouraged their efforts in constructing things of their own designs. "Well do I remember submitting to her our plans for our first flying machine, to the construction of which she readily consented. Less encouraging was an apprehensive uncle, who constantly prophesied disaster," Gustav recounted in an introduction to a posthumously published edition of his brother Otto's *Birdflight as the Basis of Aviation*. Because in 1873 there was no German aeronautical society, Gustav and Otto Lilienthal joined the British one. They worked at first on designing and constructing flying contraptions, including some with flapping wings.

Impoverished by the early death of his father, Otto Lilienthal was eventually awarded a scholarship to attend the Technical Academy in Berlin, through the efforts of its Director, Franz Reuleaux. Wittgenstein would later go to Berlin to obtain an engineering certificate. Biographers and scholars of Wittgenstein have not, so far as I know, suggested a connection between Wittgenstein's choice of an institution at which to obtain an engineering certificate and the institution or city at which Lilienthal studied engineering and conducted his gliding experiments. However, most mention Wittgenstein's interest in aeronautical engineering. Memorials and obituaries of Lilienthal portraying him as a heroic and scientifically sound investigator of the secrets of flight were plentiful during Wittgenstein's boyhood. Therefore, the information that Lilienthal studied engineering in Charlottenberg near Berlin and that his mentor was Franz Reuleaux was likely to be remembered when it came time for Wittgenstein to decide on an educational institution after the Realschule.

Gustav and Otto were separated at various times during their lives, but when Otto was at the Technical Academy of Berlin, Gustav moved there to join him. By living simply, they were able to use some of the money from the scholarship to purchase the supplies they needed to build flying machines. Gustav moved around, living in Austria, and then in England, but he rejoined Otto in Berlin again, after which they built machines with steam-powered engines attached. He also reported on some experiments with kites, one of which had an especially interesting result (the third person of the three holding kite lines in the experiment described here was his sister):

> We also built kites in the form of birds, in order to study the behaviour of the apparatus in the wind; the surfaces of the wings being curved, in order to imitate a bird. Such a kite, which we flew on the high plain between the Spandau Road and the railway to Hamburg, showed some peculiar properties. It was held by three persons, one of whom took hold of the two lines which were fastened to the front cane and to the tail, respectively, whilst the other two persons each held the line which was fastened to each wing. In this way it was possible to regulate the floating kite, as regards its two axes. Once ... during a very strong wind, we were able to so direct the kite that it moved against the wind. As soon as its long axis was approximately horizontal the kite did not come down but moved forward at the same level.

During a time when Gustav was in Australia, and the brothers were apart and hence not performing experiments together, Gustav reports that (as the Wright Brothers later related doing as well) he had tried to "construct on a larger scale the small flyers moved by a rubber spring"—with only negative results. After Otto had settled into a stable profession and built a house with a lawn and a laboratory, Gustav joined him once again, and they embarked on the more organized experiments reported in Otto's book and, finally, on the human-carrying gliders. In 1894, they had a special hill built near Berlin to give them a suitable place from which to start their gliding flights. They also found some bare sandhills near Stoellen that were excellent for gliding. They had planned to remove their gliders from Stoellen on a particular August Sunday in 1896, but, Gustav reports, he had an accident with his cycle and could not accompany Otto to the sandhill. Otto went on without him and went gliding. Doing so while separated from Gustav was very unusual. This was the flight that ended tragically in a crash. Otto died soon afterward from the injuries he received. Hence, Otto had the terrible accident that ended his life while he was gliding without his brother present. Before dying, he uttered the phrase that would be put on his gravestone and often repeated: "Sacrifices must be made." But it is not clear that *that* sacrifice had to be made; Gustav attributed the injuries Otto received from the accident to Otto's neglect to fit the shock absorber apparatus normally used as a safety measure.

The relationship between Gustav and Otto must have seemed familiar to Wilbur and Orville Wright—except that it was Wilbur, the elder, who seemed worried that Orville would not take the proper precautions without him there. It was a worry that was later justified when Orville crashed and was seriously injured while Wilbur was away in Europe. They had other things in common, too. Both sets of brothers started and ran various business ventures together. As mentioned earlier, both had mechanically minded mothers who encouraged their activities and could give practical mechanical advice. And each set of brothers also had a sister who took an interest in their flying experiments and with whom they lived at various times as adults. There is something a bit eerie about the work of the Lilienthal brothers being picked up and carried on to completion by another set of brothers in America after Otto's death. Gustav writes that their father died just after he had prepared to emigrate to America, so we get the sense that the Lilienthal Brothers' experiments might just as well have occurred in the U.S., and their gliding flights might have been carried out on sandhills in North America instead of near Berlin, but for the change of course caused by their father's death.

It was more than fifteen years after his first lecture in 1873 on the theory of bird flight that Otto's book, in which he lay out his reasoning, his observations, and his experiments, was published. The Lilienthals did not start out testing a human-carrying apparatus as so many other "inventors" had. In his book, Otto explains how and why he first laid out a research program based on observations of various types of flying creatures, built a test apparatus with which to test model wings (a whirling arm, shown in a woodcut in the book), designed and recorded measurements from an organized series of experiments in which he measured the force of air on surfaces of various shapes and at various inclinations, and performed detailed calculations to explain observations and predict performances using energy considerations. The book was written in an engaging style and illustrated with many woodcuts. It was accessible to nonspecialists, though it contained the substantial technical content of his thinking and the upshot (including woodcuts of airflow patterns) of his experimental results. Lilienthal's book was well known and was probably easily had in Vienna. The possibility of human flight interested not only adults keeping tabs on technology; young people and children were also fascinated by the thought that humans could fly. Lilienthal wrote the book as if investigating from scratch, and he begins by inviting the reader to first shed any preconceptions about how birds fly and to instead follow along with him on his observations.

There is in his description of this stage of human knowledge an image of the human situation that was later evoked in describing the human condition with respect to language and logic: the image of humans who can move on the surface of the Earth in any direction, on both land or sea, without coming to the end of the world, yet are bound to the surface of the Earth. In this image, humans are aware that there is some other place outside the limits of their range of movement, but they are not familiar with it nor able to explore it in the same way as they can places within the limits of their range of movement. They cannot move about in the atmosphere at will, as they can on the surface of the Earth.

Lilienthal asked wistfully about the flight of birds: "Are we still to be debarred from calling this art our own, and are we only to look up longingly to inferior creatures who describe their beautiful paths in the blue of the sky?" And, interestingly, in his book he separated the question of *being able to fly* from that of *knowing how flying is done*:

> Is this painful consideration to be still further intensified by the conviction that we shall never be able to discover the flying methods of the birds? Or will it be within the scope of the human mind to fathom those means which will be a substitute for what Nature has denied us? ... It must not remain our desire only to acquire the art of the bird, nay, it is our duty not to rest until we have attained to a perfect scientific conception of the problem of flight, even though as the result of such endeavours we arrive at the conclusion that we shall never be able to transfer our highway to the air.

It is also interesting that what scientific understanding might enable humans to do is not quite the same as what the birds do naturally: "... it may also be that our investigations will teach us to artificially imitate what nature demonstrates to us daily in birdflight." There is another interesting parallel between language and flight. Just as people were soon to speak of artificial versus natural flight, so there would soon be in Europe debates about the possibilities and virtues of artificial language (such as the Esperanto movement and variations on it, or formal mathematical systems) versus natural language in the intellectual atmosphere surrounding young Ludwig Wittgenstein.

Lilienthal's investigation begins by noticing that the principle explaining how birds are able to fly must be a very different principle from the principle that explains why balloons stay aloft in the air. Any principle of bird flight, Lilienthal infers, must be a principle "in which only very thin surfaces come into consideration, which offer very little resistance to the air in passing through it in a horizontal direction." After his experimentation with plane (flat) surfaces, he determined that plane surfaces are incapable of providing the lift needed for human flight, no matter what the angle of inclination, and he renewed the appeal to "Nature":

> Nature demonstrates daily that flight is not at all difficult, and whenever we feel inclined to give up the idea of human flight, owing to the apparently unattainable energy required, every large bird passing us with slow deliberate wing-beats, every sailing swallow rekindles in us the thought that our calculations cannot be correct ... that somewhere there is a hidden secret which, once disclosed, would completely solve the mystery of flight.

Some scholars see this kind of appeal to nature (appeals of the form "there are organisms in nature that exhibit a certain thing; hence, the world must be of such a sort that that certain thing is possible") in Wittgenstein's remarks on the use of symbolism. That is, they see in the *Tractatus* appeals to the fact that language (being able to use symbols to mean things) is something human organisms are capable of. Hence (whether or not one has a theory of language or symbolism), the world must be of such a sort that language is possible.

It was important to Lilienthal to distinguish the principle responsible for balloon flight from the principle responsible for bird flight. This was important for him in order to lay out his own research program, of course, but he also felt it was important that it be more generally understood as well. He felt that the discovery of the balloon as a means of humans' being able to stay up in the air had impeded the development of mechanical flight. His argument was that people were easily misled into thinking that the major problem of flight had been overcome with the advent of lighter-than-air aircraft, and that only the more minor problem of steering a balloon remained: "It must have been intoxicating, when a century ago the first man actually rose from the earth into the air; [...] What satisfaction it was, after striving for 10,000 years, that the ocean of air had opened its unlimited space to us!" This intoxication had its deleterious effects on perception and judgment, however: "It could not be difficult to utilize this new element for free locomotion, and it seemed as though it required very little, a mere detail, in order to solve the great problem of aerial navigation." It turned out this "mere detail" was instead an "almost insurmountable difficulty." Hence, it was becoming clear that "the balloon will remain what it is, namely, a means of rising in the air, but no means for practical and free navigation of the air." In the meantime, however, Lilienthal argued, a hundred years had been lost, as flight researchers had been distracted by the problems of steering a balloon instead of investigating the principles by which birds fly. He even referred to the balloon as a "direct brake" on the development of dynamic flight.

Despite the ill-directed and wasted efforts spent on the hopeless task of making the balloon steerable (dirigible), and Lilienthal's judgment that the discovery of the balloon for aerial observation was "a setback," the aerial balloon was not completely useless, even in Lilienthal's view. He may have argued that it will always be "what it is," but he was willing to grant that "what it is" had its uses: as "an elevated observation post in its captive form" or a form of travel in which one is "wafted along by the wind." What Lilienthal was urging was that the research efforts then being directed toward making the balloon dirigible (steerable) instead be directed toward the goal of being "able to travel through the air rapidly and safely in whatever direction we desire, and not only in the direction in which the wind blows."

In taking on ballooning like this, Lilienthal was up against some powerful images and romantic associations. In comparison to the awe-inspiring grandeur of balloons used for pleasure travel and the subsequent practical development of the balloon for military transport and operations, the kite was a rather humble object, often regarded as merely a toy. It gradually gained some respect with the news of Hargrave and Cody's applications of large box-style kites. The history of ballooning, on the other hand, had been as exciting in the preceding century as making "flying to the moon" a reality would be in the next.

As Hallion tells the story of the Montgolfiers' invention, Joseph Montgolfier came upon the idea while pondering the great military advantage a balloon would provide in sieges, although he was not the first to conceive of the idea of using warmed air in an enclosed space so as to cause a container to rise. Joseph soon interested and enlisted the help of his brother Étienne in designing, building, and experimenting with various kinds of enclosures for the warmed air. They tried to get the French Academy interested, suggesting balloons might be used for military transport, bombing operations, and scientific investigation of the atmosphere, but nothing came of it at first. They persisted on their own, with limited funds, and their subsequent public demonstration of a 35-foot-diameter balloon in 1783 was "a spectacular success," astonishing the onlookers. Then, as Hallion puts it, "a curious 'balloon race' began, pitting the craft tradition of the Montgolfiers against the physicists and chemists of the French and British scientific establishment." The development in ballooning of a contest between visionaries equipped with only their own insight, curiosity, and talent against scientists with prestigious reputations and access to more extravagant resources foreshadowed the future of aviation, in that something similar would occur in the development of heavier-than-air flight.

In the case of ballooning, it was J. A. C. Charles, a member of the French Academy of Science and a friend of Benjamin Franklin (the U.S. ambassador to France), who was from the scientific establishment. Located in Paris, and well situated for obtaining the requisite funds, he had access to large amounts of a gas that was lighter than air and so required neither initial nor continued heating to keep the balloon aloft. Being only about a tenth the weight of air, Charles's balloon did not require nearly the volume that hot-air balloons did. In late August of that same year, just a few months after the Montgolfiers' spectacular display in Annonay, Charles (assisted by yet another set of brothers, Jean and Noel Robert) successfully flew a 12-foot-diameter balloon over Paris.

That individual balloon, called a "globe," came to a tragic end: the rural residents of the hamlet in the countryside surrounding Paris where the globe landed literally butchered the frightening but fragile thing. But the hydrogen balloon as a new type of balloon had got its start in spite of the fate of that individual globe, for the use of hydrogen rather than warmed air was a technological advantage, and Charles' variation on the Montgolfiers' invention did prevail. But the Montgolfiers were still recognized as introducing this astounding new possibility, in part because they were the first to put humans into flight, and in part because they made their balloons objects of charm and visual delight. Their balloon was gargantuan, but "as gaily decorated as a Faberge egg: royal blue with gold trim showing the twelve signs of the Zodiac, the king's initials, the head of Minerva in a radiant sun, and *beaucoup* banding and entwined filigree, including a repetitive pattern of fleur-de-lis."

It was not only its visually appealing appearance to onlookers that made the balloon such an object of delight. As Lilienthal later surmised, human air travel by balloon was, by all firsthand accounts, intoxicating. Besides the exhilaration of floating and being surrounded by the sky on all sides, air travel provided something of a totally different sort: a view of the world from without. One could view the Earth without standing anywhere on the Earth. In contrast, the view from even the highest mountaintop, though providing an aerial view of large expanses of the Earth from a vantage point among the clouds, did not include an aerial view of the terrain on which the viewer stood. From a balloon, one could view the Earth as a celestial object.

It did not take long for the feat to become appreciated. With the advent of the Montgolfier, human flight (in the sense of being able to travel in the air) was no longer a wistfully longed-for dream of the imagination. While still potentially dangerous and experimental, human flight had, by the end of 1873, been successfully carried out and witnessed by many tens of thousands of people. It had happened so fast—in barely the course of a year from the first hint of the possibility of air travel by balloon to its recognition by a stunned and delighted public. Balloons were talked about everywhere, and balloons and aeronauts were depicted on dinnerware, clothing, clocks, and jewelry. Hallion describes the exasperation some felt at such an all-pervasive mania for the balloon, such as when Samuel Johnson wrote to a correspondent: "I have three letters this day, all about the balloon. I could have been content with one. Do not write about the balloon, whatever else you may think proper to say."

In the century that followed, balloons were developed for military use by many countries (such as for aerial bombing, aerial reconnaissance, communications, and transport). Nevertheless, the problem of steerability remained a serious limitation. There was just no getting around the fact that, due to the large surface area of a balloon exposed to the wind, the wind could overwhelm most other means of directional control. The Austrian Army's failed attempt to use balloons for bombing had illustrated this strikingly:

> The first aerial bombing was attempted in 1849 when the Austrians launched 200 pilotless, bomb-carrying hot-air balloons against forces defending Venice. Each bomb was released by a time fuse. However, the wind sent the balloons back over the Austrian troops.

Even with human pilots, and later improvements in powering and controlling balloons, human air travel by balloon was not anything like flying like a bird; it was more like floating like a cloud. Yet the love affair with balloons did not end, at least not for the general public. Even a dozen years after Lilienthal wrote his book, Alberto Santos-Dumont's powered cigar-shaped balloons were winning him prizes, personal celebrity, and public affection. There continued to be refinements in designs of dirigibles designed for military use.

Among serious investigators into flight for its own sake, however, there was a growing appreciation of the kite and propelled model aeroplanes in spite of the balloon's popularity. The spectacular public success of Pénaud's model aeroplane in 1871 was enough to inspire conviction in the possibility of practical heavier-than-air flight and to show what sort of craft might be able to achieve it. Lilienthal's lecture in 1873 was followed by a number of like-minded efforts. The French-born American engineer Octave Chanute began collecting and reading whatever he could find on the subject of experiments with heavier-than-air flight around the same time, considering it a sort of forbidden indulgence of time and attention. As it became less a sign of insanity and more acceptable as a hobby, Chanute eventually summarized his extensive literature research in a series of articles (published in *The Railroad and Engineering Journal* from 1891 to 1893) that were subsequently published as the book *Progress in Flying Machines*. In the U.S., too, Samuel Langley began his model aeroplane experiments around 1887, publishing the results of his tests of powered models.

Almost a decade after Lilienthal's book *Der Vogelflug als Grundlage der Fliegekunst* (*Birdflight as the Basis of Aviation*) was published in Berlin in 1889, the American-born Hiram Maxim, who was living in England, published a book called *Natural and Artificial Flight*, giving an account of his experiments up until then. It was similar to Lilienthal's book in some ways, making the same points about the possibilities and superior potential of heavier-than-air flying machines over lighter-than-air craft. Maxim reported that in 1889 (again, the year of Wittgenstein's birth!), he carried out experiments with full-scale versions of a flying machine powered by a steam engine. Maxim, too, emphasized the value of the kite for research into heavier-than-air flying machines, but he was especially interested in showing that it was possible to design an engine that was both sufficiently lightweight and sufficiently powerful to propel a flying machine to the appropriate speed, such that it would go fast enough that the lift generated by the surfaces of the kite-like structure of the machine would counteract the weight of both the machine and the engine.

Maxim's work was striking in comparison with the other kite researchers and model aeroplane designers of the time in that it was a full-scale powered machine meant to carry human passengers. Because he could not arrange to fly the machine in Baldwyn's Park on satisfactory terms, he constructed a pair of rails that would constrain the device from rising should the lift force developed be sufficient to raise it into the air. As with so many other inventors and flight researchers, Maxim had already made a fortune in something else—he had invented the automatic machine gun before moving to England from America. Maxim was good at publicizing his efforts in flight. Before writing the book *Natural and Artificial Flight*, he had described his preliminary model experiments in periodicals such as *The Century Magazine* and *Manufacturer and Builder*, and had given interviews about them. Maxim stressed the political and military consequences of his achievement, too.

Cody and the Wrights had thought of military uses for their inventions too, but whereas they had thought of using kites and airplanes for observation, Maxim had more brutal uses in mind. The Wrights had, perhaps naively, even thought that the advent of the airplane would make war obsolete, as the improved aerial observation the airplane would provide would make any preparations for attack visible and hence ineffectual. Cody's application of the "Man-Lifting War Kites," too, was for observation of enemy troop movements. Maxim had invented the automatic-firing machine gun, which, depending on one's viewpoint, was striking for its fantastic efficiency or ruthless brutality. *Manufacturer and Builder* portrayed him from the latter viewpoint:

> Mr. Maxim is reported to have opened the interview with the following bloodthirsty announcement: "If I can rise from the coast of France, sail through the air across the Channel, and drop half a ton of nitro-glycerine upon an English city, I can revolutionize the world. I believe I can do it if I live long enough. If I die, someone will come after me who will be successful where I failed.

Perhaps he had in mind the previous failed Austrian attempts at using balloons for bombing. He certainly did mean to contrast heavier-than-air flight with balloons. His emphasis was not on steering devices or control, though, but on the engine, and on how much horsepower it took to get the machine to achieve a certain speed. Before Maxim had exhibited control in flight (which is what Lilienthal and the Wright Brothers considered the first step), he was working on going fast. This made some sense, since, as was well known by then, going faster resulted in more lift. Like Lilienthal, Maxim argued for the importance of research into heavier-than-air flight, pointing out the lack of control inherent in flying lighter-than-air craft. Even when he complained that travel by balloon was "at the mercy of the wind," however, his concern in designing heavier-than-air craft was in achieving speed and overcoming the sheer force of the wind, with very little appreciation of problems that could arise in control of heavier-than-air craft. Perhaps this reflects that he was unwittingly making the same error with respect to heavier-than-air flight that many had made about lighter-than-air flight: thinking that controlling the aircraft would be a detail, and hardly something significant enough to dampen the exuberant celebration of becoming airborne.

In the interview reported in the 1891 *Manufacturer and Builder* article, Maxim emphasized the *difference* between his technique and nature's means. This is a somewhat different comparison than Lilienthal and Verne's fictional hero Robur had drawn, for they had pointed out that heavier-than-air flight (rather than lighter-than-air flight) was nature's means of aerial transport of large animals, and so that the airplane was *similar* to natural means of aerial transport. The Maxim interview took place several years prior to Lilienthal's death. In fact, Maxim may have had Lilienthal in mind when he said that it is not necessary for a flying machine to copy the wing-flapping of birds. Some of Lilienthal's early machines did employ wing-flapping, although Lilienthal was well aware of the ability of properly shaped stationary wings moving with respect to air to provide lift without requiring wing-flapping mechanisms. Lilienthal, in return, was critical of Maxim's efforts toward building a flying machine. In his biography of the Maxim brothers, McCallum notes that:

> The distinction made by [H.G. Wells in *The Argonauts of the Air,* published in 1895] between the effort involved in taking off and the skills required to control the machine in the air was of course a crucial one. It had already been emphasised by Chanute and by the German Otto Lilienthal, who was succeeding with ever more ambitious flights in his hang gliders and was critical of Hiram's efforts, declaring that the American was on the wrong track and that even if his machine had taken off he could not have maintained it in flight. This was true, and indeed Hiram was the first to admit that much practice would be needed before he could expect to achieve such control. For his part he affected to dismiss Lilienthal as a mere parachutist, likening him to a 'flying squirrel,' and the German responded in kind, retorting that the one achievement of Hiram's massive aeroplane had been to show others how not to fly.

The contrast Maxim drew was an indication of his emphasis on a means of propulsion rather than on a means of control. Maxim was advocating a screw propeller. He argued that, just as the wheel was a better mechanical mechanism for travel on land than a mechanism designed to imitate the walking mechanism for travel on land found in nature in animals such as horses, so the screw propeller was superior to wing-flapping. Like many others around that time, he experimented with kites, and his reasoning about how to design a flying machine began with reflecting on his experiences with flying a kite:

> As is well known, when one flies a kite the cord holds the kite against the wind. The wind power passing on the under side of the kite strikes it at an angle and raises the kite into the air. [...] If the angle be slight, the amount of strain on the cord necessary to hold it against the wind will be found considerably less than the weight of the kite and the load which it is able to lift, particularly so if the cord pulls in a horizontal direction instead of at an angle. It is also well known that if a kite be propelled in a calm through the air, ... the effect is exactly the same. Suppose now, instead of the cord for holding the kite against the wind or for propelling it against still air, that a screw propeller should be attached to the kite and that it should be driven by some motor. If the screw propeller could be made to give a push equal to the pull of the kite, and if the machinery for driving it should be no greater than the weight that the kite would be able to carry, we should have a veritable flying machine.

Maxim wanted to go fast, really fast, and he was good at designing light, powerful machines. He wrote about the trial of his lightweight steam engine-powered machine at Baldwyn's Park, describing how it first went 30 miles an hour, then 35, at which it "began to rise," and then how it went 90 miles per hour, at which "it pulled its guy wires with such force that it broke them, and now we have to keep it chained." Chained—like a powerful animal capable of destruction. In discussing artificial and natural flight, Maxim often argued for the propeller rather than wing-flapping mechanisms by alluding to the superior capability for speed that the locomotive had over the horse. Comparing locomotion by wheels with locomotion by legs was not unfamiliar: during a phase of his life spent in Europe, Cody had entertained all of Paris with his challenges to cyclists that he could beat them on horseback. The two different kinds of comparisons made between machine design and nature that arose in advocating heavier-than-air flight are interesting because they provide contemporary parallels between kinds of artificiality in machine design and kinds of artificiality in languages.

The lives of many of the eccentric and inspired people involved in early flight research and invention were full of adventure and drama. Like so many other flight pioneers, Hiram S. Maxim was part of a team of brothers who worked together. Their relationship, however, became stressed and eventually fractured from mutual resentment, accusations of thievery of patents, and the spectre of blackmail. This is only one of many such interesting facets of their lives. Iain McCallum made their relationship the subject of an entire book: *Blood Brothers: Hiram and Hudson Maxim—Pioneers of Modern Warfare.*

Maxim kept painstakingly detailed records of his work, and he had his apparatus fitted with instruments that would record the forces in the springs that held the machine captive to the Earth. He sought and enjoyed an audience, and he announced some of the "trials" he made of his machine, sometimes holding special by-invitation-only events. One such event in 1894 was held for the leading scientists in England. Though it met with mixed reviews due to the mishaps that occurred from which Maxim barely escaped serious injury, the attention of distinguished guests at least ensured that his efforts would be noticed, and it was discussed widely in newspapers and journals. The main claim Maxim made for the event was that it represented a milestone in flight: the raising into the air of a machine carrying a human by its own mechanical power (his steam engine).

Ludwig Boltzmann heard about it in Vienna. Although he was excited about the achievement it indicated, it also made Boltzmann a bit hopeful about the possibilities for Germany to succeed where England had not. This is when he gave the lecture "On Aeronautics" in Vienna on the topic. I have mentioned Boltzmann's importance to Wittgenstein, who had hoped to study with him as a teenager, admired him all his life, and explicitly cited him as important in his intellectual development. Boltzmann had this personal significance to Wittgenstein and was a looming figure to others, too, in Wittgenstein's boyhood and adolescence, due to Boltzmann's very public presence in Vienna. He was also one of Europe's preeminent figures in science, a beloved and famed scientist who wrote a textbook on mechanics that discussed philosophical and foundational issues, and he was working on thermodynamics. His approach was generally viewed as a contrast to Wilhelm Ostwald's, for he was developing the kinetic theory of gases. In this theory, macroscopic phenomena such as a gas's temperature and pressure are explained in terms of the motions of particles that are themselves too small to be observed. He was one of the theory's most vocal proponents, and he was known for the public lectures he gave in Vienna arguing for the "atomic hypothesis." As a major figure in the fields of science and philosophy, his views on the controversial topic of the achievability of heavier-than-air flight were sought. To him, as to many, it seems, the question of which invention—or even which individual's invention—would first succeed in practical heavier-than-air human flight was secondary to the question of which country would be able to claim the honor.

Boltzmann was insightful enough to realize that Maxim's much-publicized demonstration to an audience of scientists did not signify a solution to the problem of heavier-than-air flight, but he did think the event was a milestone of sorts. He used it to stir interest and call for German investment in flight, in the lecture "On Aeronautics" that same year, 1894, delivered before the *Gesellschaft Deutscher Naturforscher und Ärzte* in Vienna. (As I have mentioned, it was reprinted in his 1905 collection *Popular Writings*, which coincided with the period during which Wittgenstein formed an intention to study with him. A translation of the lecture into English produced especially for this book is included in an appendix; so far as I know, no other English translation is available.)

Many of the points that appear in Boltzmann's lecture were already much discussed: the limitations of balloons, the promising examples of heavier-than-air flight provided by kites and large birds, the comparisons of machines with nature's means of accomplishing the function, the importance of solving the problem of control. The essay was not meant as an original contribution to the science of flight, though. It was, rather, an observation and analysis of where things stood, with some inspirational remarks about the work yet to be done.

Boltzmann analyzes the problem of flight on two levels. On one level, he discusses it as an age-old problem once thought insoluble but on which enough progress on questions relevant to it had lately been made that it now appeared not only soluble, but ripe for solution. He analyzes exactly what solving the problem would consist of and what sorts of skills and character traits are called for to do so. On another level, he analyzes the problem of flight as a technical problem: he describes the technical accomplishments of late that are steps toward a solution and gives his opinions as to which means of propulsion he feels will work.

Boltzmann's discussion of flight as an age-old problem ripe for solution is almost poetic. He uses Gauss's fifth problem, known as the Kreisteilung (division of the circle, or circle-partitioning) Problem as a metaphor. Carl Friedrich Gauss had a position at Gottingen as director of the Observatory; when he died, a few days after Boltzmann's eleventh birthday, he was one of the most famous mathematicians in Europe. He had won a major prize for a work that included "the idea of mapping one surface onto another so that the two are similar in their smallest parts." The Kreisteilung Problem was very significant to Gauss personally: it established him as a mathematician of unusual talent very early in his career. The Kreisteilung Problem—the problem of deciding whether a circle can be divided into n equal parts using only ruler and compass—had its roots in ancient geometry. But Gauss did not investigate it as a geometrical problem; he was investigating solutions to the equation $x^p = 1$, which are known as "roots of unity." These are solutions where x can be a complex number. For example, the complex number i defined such that $i^2 = -1$ is a fourth root of 1, since $i^4 = 1$. Thus, the complex number i is a "root of unity." There is a geometrical representation whereby this fact is represented by showing that i divides the circle into four equal parts. In fact, any root of unity can be represented geometrically as dividing the circle into an equal number of parts.

A further question is which of these roots corresponding to dividing the circle into equal parts—into a regular-sided polygon—is "constructible" using only ruler-and-compass methods. Gauss figured this out, and he figured it out because he was investigating and categorizing solutions to the equation $x^p = 1$. He determined criteria for the solutions to this equation for which the corresponding polygon would be constructible. He established that a regular 17-sided polygon was constructible. More generally, he showed that a sufficient condition for the constructibility of regular polygons is that the number of sides be a Fermat prime. Fermat primes are numbers of the form $Fn = 2^{2^n} + 1$, where n is a nonnegative integer and the number Fn is prime. So far, the only known Fermat primes are F0, F1, F2, F3, and F4.

The metaphor works in several ways. One of them is that the problem of flight was then, as the Kreisteilung Problem had been in Gauss's youth, a historically significant problem ripe for an unknown individual to apply his genius to and establish a reputation. Another is that the problem of flight until recently was, as the Kreistielung Problem had been prior to Gauss's time, a beguiling age-old problem whose solution was attempted unsuccessfully by so many that it was considered unsolvable. Boltzman made the comparison explicitly:

> ... in the same way the solution of the problem of the division of the circle (Kreisteilung) failed before Gauß, the construction of the dirigible failed, so that the problem became seriously discredited. Yes, great theoreticians were actually inclined to believe that its solution was impossible. Only recently a turn occurred. The inaccuracy of the old formulas has clearly been shown, and I believe I can provide a proof that the solution to this problem is not only possible but will in all likelihood be found shortly.

Boltzmann goes on to explain that his proof will not be a proof of theoretical mechanics, which is helpless to solve the problem of flight, due to the complexity of the problem, but that he will "state the leading ideas and point out fundamental terms." Here is where the metaphor works in yet another way: after laying out the basic approaches of balloon, helicopter, and aeroplane and endorsing the last of these, Boltzmann discusses means of powering an aeroplane. He prefaces the list of types of machines that might be considered for the purpose with the remark that "All mechanisms that are used in technology create a so-called cyclic motion, i.e. a motion in which all constituent parts will again arrive at the starting-point after a shorter or longer amount of time." Of the two kinds of cyclic mechanisms—rotating and reciprocal—Boltzmann argues that the rotating will be the one found to produce a steerable heavier-than-air flying machine. Characterizing mechanisms used in technology in terms of what kind of cyclic mechanism they are rather poetically invokes the geometrical representation of roots of the equation $x^p = 1$. One can represent the roots of the equation in terms of a point traveling around the circumference of a circle, taking on values of the roots of the equation $x^p = 1$ and then return-ing to its original position. Repeated multiplication of 1 by an nth root of unity takes it from the starting point of 1 to each of the n roots of unity on the circle in succession, and then around the circle again and again.

But with the rotational movement of the airscrew mentally overlaid on this geometrical representation of the roots of the equation $x^p = 1$, the metaphor appears to end. What could it mean for solving the problem of practical flight that the answer Gauss found for the Kreisteilung problem was stated in terms of a kind of number, the Fermat prime? A prime num-ber is simple in that it has no factors other than itself and 1. Is there any-thing about flying machines like what Gauss showed: that whether or not a geometric figure of n sides could be constructed was reflected in the fact that when n appeared in a certain equation it ensured that that equation could be reduced to a simpler form? And what could this metaphor imply for solving the problem of practical flight, given that Gauss found the answer to the geometric problem of the constructibility of a certain kind of polygon by investigating the complex (real and imaginary) roots of an algebraic equation?

Although several of Wittgenstein's biographers have mentioned Boltzmann's lecture on aeronautics in passing, it has not been read much. The lecture was given in Vienna in 1894, when Wittgenstein was five years old. He was then living in a grand house in the cultural center of Vienna, in a household that was full of older siblings who kept abreast of the city's intellectual life and that was headed by a father who kept abreast of trends in technology. Perhaps even more relevant to the possible effect a lecture about an as-yet unknown hero who would solve the problem of flight might have had on a young child, Boltzmann's lecture mentioned the sig-nificant insights to be gained from experimenting with toys.

In discussing what he calls "dynamic flying machines," which he con-trasts with balloons, he says:

> They fall into two main categories. In the first category, the motive power is preferably used to lift the device. For this purpose usually one or more airscrews are used which revolve vertically upwards, just like the screw of a steamship revolves horizontally in water. As in that

case, a small part of the whole surface of the screw—two or four uniformly inclined planes that revolve forward in rapid rotation by virtue of their inclination—suffices. The model of this apparatus is a well known children's toy.

This toy is presumably the Pénaud helicopter, or some imitation of it. He goes on to say: "Imagine two or four of such enormous airscrews which are turned very rapidly by a machine: if mounted on a heavy object, it can be lifted up in the air; and you have the helicopter." However, the configuration he thinks most promising is to use the airscrews horizontally to propel an aeroplane forward. Here, a completely different kind of toy illustrates the principle involved: a kite. He says:

> In contrast to that, in the second class of the dynamic flying machines, the gliders or äeroplanes, the motive power is mainly used for horizontal motion. Lift-off comes about—according to the most accurate measuring principle followed by Wellner und Lilienthal—due to using a plane with small inclination and curvature which is then … drawn through the air at rapid speed. We will call it the principle of the inclined plane. This principle can also be exemplified by a well known toy—the kite.

He then relates the news of Maxim's awe-inspiring trial of the huge 8,000-pound machine that has to be chained to be kept from escaping into the sky:

> … Maxim has decidedly made the second big step towards the invention of the dirigible. He has shown that one is indeed able to lift huge loads up into the air by means of a dynamic flying apparatus. The greatest English physicists, who are all theoreticians—Lord Kelvin, Lord Rayleigh, Lodge, etc.—spoke of Maxim's machine with excitement.

But this is not the solution to the problem, Boltzmann explains; it is only a precursor to it, an indication that the problem is soluble. What would constitute solving the problem would be being able to fly in an arbitrary direction—that is, with the wind or against it, as one wished, for a sustained period of time—say, an hour. It is the German Lilienthal, he said, who has begun on the step that will lead to solving the problem, and he is doing so on a much smaller budget than Maxim's grandiose efforts required. Boltzmann implies that Lilienthal is doing all the right things: learning how to control a glider and using a small one to learn what is involved before trying to control a large one, where the destabilizing forces are so much larger. Lilienthal didn't even have a motor on his current gliders; he used his "jump to flight" method, jumping from an elevated spot. In this way he learned to control his flying machine, and so to sustain his flight for longer and longer distances. Though it would take a much larger flying machine to have major economic and social consequences, Boltzmann said, a small mechanically powered flying machine big enough to carry a human would, if it were capable of controlled sustained flight, mean the problem had been theoretically solved.

After describing the characteristics of the flying machine that he predicted would achieve success, Boltzmann described the characteristics of the person who would invent and successfully pilot a flying machine:

> Only he who possesses the courage to trust his life to the new element, and the cunning to overcome gradually all its treacheries has a chance to kill the dragon which, until this very day, deprives mankind of the treasure of this invention.

As a child of five years of age, Wittgenstein might have been able to grasp at least this much from what he heard of or about Boltzmann's lecture: the next hero would be the person who first figured out how to control a kite, and it didn't even have to be a huge kite. It would eventually have to be powered, but that was something a child could work with, too— that meant using a propeller, just like the propeller on a Pénaud-style helicopter toy already familiar to children all over Europe. Having these toys and working hard to figure out how to make them do as you wished was not enough, though. It would take a genius, and a genius really serious about the problem, too—a genius so dedicated to solving the problem that he would risk his life for it. Lots and lots of people had tried to solve this problem. It was worth trying only if you were the rare genius, like the mathematician Gauss, who would solve it.

Boltzmann's lecture, delivered in Vienna in 1894 when Ludwig Wittgenstein was five years old and living there, was later reprinted in Boltzmann's *Popular Writings*, which was published in Berlin in 1905 when

Wittgenstein was sixteen. As we shall see, by the age of sixteen, he had become an earnest youth interested in aeronautical engineering as well as in physics —and the person he especially wanted to study with was Ludwig Boltzmann.

Chapter 3

Finding a Place in the World

When Boltzmann delivered his lecture in Vienna in 1894 alerting the technical community that the time was right for the right kind of person to solve the problem of practical flight, lauded Lilienthal's approach, and explained how the major principle to be investigated—the "principle of the inclined plane"—was exhibited in a well-known toy, Wilbur and Orville Wright, in the American Midwest, had not forgotten about the Pénaud-style helicopter toy they had been given as boys. The original was long destroyed by then, but they were trying to build larger versions of it that worked as well as the tiny toy. In 1894, neither Wilbur nor Orville Wright was a child anymore, as Ludwig Wittgenstein was.

Yet the Wrights, although settled into a comfortable lifestyle operating the custom bicycle manufacturing business they had recently started, were still finding their way about the world. They were still seeking out information about whole new worlds of knowledge they were unfamiliar with, still looking for something more satisfying to do. They certainly didn't need to find a new profession, for they had plenty to occupy their time. They were instead enticed by the news of the unsolved problem that had intrigued them as children. Their pursuit had something of the passionate character of an obsession, yet they proceeded methodically and took their time to do things thoroughly.

There is a parallel between the Wrights staking out their place in the history of aviation and Wittgenstein staking out his in the history of philosophy; before the accomplishments for which they are now famous, they were already well positioned in other avocations. Just as the Wrights rapidly learned as much as they could from the mentor they sought out (Octave Chanute) solely for his knowledge about the problem in flight they hoped to solve, so Wittgenstein sought out mentors (Frege and Russell) in the problems in logic he wanted to solve, and he rapidly learned as much as he could from them. The Wright Brothers then went beyond what they had learned from Chanute, on paths they struck out for themselves, just as Wittgenstein went beyond what he had learned from Frege and Russell, eventually going off on his own, on a path he struck out for himself. Still, Chanute's reputation and standing in the world of aeronautical researchers gave the Wright Brothers credibility and entry to audiences they would not have had otherwise, and Chanute encouraged them to continue when they were frustrated and losing confidence. Russell played a similar role in making Wittgenstein known to a larger audience, first within Cambridge, and later the world, and in encouraging Wittgenstein's development as a philosopher when he needed assurance that he should continue in philosophy. In the end, though, the Wright Brothers felt Chanute did not really appreciate the significance and uniqueness of the contribution they had made, and Wittgenstein felt that Russell and Frege did not understand the work he produced that became the *Tractatus*.

In 1894, Wittgenstein was a young child and still had ahead of him decisions that would lead him into his first avocation; the Wright Brothers were already settled in theirs and were about to embark on the decisions that would lead to their second. Wittgenstein in Vienna and the Wright Brothers in the American Midwest found themselves immersed in an atmosphere of anticipation about the next step in heavier-than-air flight. This atmosphere was humid with the constant, dampening news of would-be inventor after would-be inventor announcing he would fly and then crashing in humiliation and injury, even death. Yet there was some refreshing air in it, too, such as Lilienthal's gliding experiments in Germany and Langley's aerodrome experiments in America. Though on opposite sides of the Atlantic Ocean, in cultures that would make for a study in contrasts, there were many commonalities in the world of Ludwig Wittgenstein and the Wright Brothers as far as what was known about flight.

One reason for commonalities in the news about flight was the activities of a generous and incredibly hard-working engineer: Octave Chanute, mentioned above. By this time he had retired, wealthy, from his career in civil engineering and was now devoting all his time to collecting and disseminating information and news about work in aviation throughout the U.S. and Europe. Ever since a vacation to France, where he had been born, he had had an enthusiasm for heavier-than-air flight, a conviction that it was possible, and the desire to find the solution himself—or at least to be part of hastening its solution. He had earlier summarized the state of the art in a series of 27 articles published in *The Railroad and Engineering Journal* from 1891 to 1893. His book, *Progress in Flying Machines*, appeared in 1894 (the year of Boltzmann's lecture on aeronautics) and was a compendium of information about the topic of flying machines that included all these articles plus one on the flight of the albatross and one titled "The carrying capacity of arched surfaces in sailing flight" by Otto Lilienthal, who was identified as "The Flying Man." Chanute also collected and disseminated detailed experimental technical information, such as the results of Lilienthal's experiments measuring lift on model wings and inclined surfaces.

The existence of aeronautical clubs and other technical societies and organizations in both the U.S. and Europe magnified Chanute's efforts to see to it that the activities of various researchers scattered geographically were made known to each other. The most significant of these were Chanute's invitations to Wilbur Wright to speak at The Western Society of Engineers—first in 1901 about the gliding experiments he and Orville had carried out, which did not bear out the predictions they expected, and then in 1903 after they had figured out the puzzling results and had a plan for a flying machine—and Chanute's own presentation about the Wright Brothers' work given in 1903 in France. Advances in printing and telegraph technology made newspapers and other periodicals an effective and timely means of communication, although they did not prevent misinformation and exaggeration, especially on such a sensational topic as heavier-than-air flight attempts.

Another source of the exchange of information about aeronautical activities back and forth across the Atlantic were international exhibitions, or World's Fairs, which played a significant role in introducing new technical developments and inventions. The one that revived the institution of the international exhibition was the 1889 World's Fair, or the Paris

Exhibition of 1889 (again, the year Wittgenstein was born!). The Eiffel Tower was constructed for the event and featured elevators. Lit with electric lights, it was a spectacular sight and a striking icon of the event as well as of the city. The Paris Exhibition of 1889 set the standard for future international exhibitions.

The 1893 World's Fair held in Chicago (also called the Columbian Exposition, as it marked the 400th anniversary of Christopher Columbus's voyage) was an extravaganza such as the U.S. had never seen. It aimed to surpass the Paris Exhibition of four years before. That such a wondrous utopian construction project could be accomplished there brought the sophistication in industry and culture of the American Midwest to the attention of the rest of the U.S., to Europe, and even to the rest of the world. It was a response to the Americans who had exclaimed of the Paris Exhibition of 1889: "Only in Paris …!" Counting repeat visitors, over 27 million attended the 1893 World's Fair in Chicago. Americans arrived by steamship, by horse carriage, and, most often, by train on the railways that ran in and out in all directions to and from Chicago. The 1893 exhibition gave many Americans their first up-close experience with new inventions such as the gramophone and electrically lighted buildings. Even the most sophisticated visitor would find a wider variety of whatever was being exhibited than he or she could have imagined. The exhibits were historical as well, often showing the development of a device, branch of technology, or cultural form from ancient times to the present.

No matter who was describing it, the Columbian Exhibition of 1893 was described as vast—just indescribably vast—and unimaginable by anyone who had not seen it with his or her own eyes. The cluster of main buildings was architecturally overwhelming: "All of the main buildings were of a uniform cornice height, geometrically logical, and covered in the same white staff (stucco), producing a homogenous yet somehow magnificent grouping of buildings." Each of the fourteen main buildings was of fantastic size, with acres and acres of floor space inside each building. In the "Palace of Mechanic Arts," for instance, were

> 79 groups and nearly 200 classes of exhibits [representing] almost every mechanical device fashioned by the ingenuity of man. There is machinery for the transmission of power, whether by electric, steam, hydraulic, or pneumatic apparatus; there are machinery and

appliances for the manufacture of textile fabrics, for the preparation
of various articles of food; for type-setting, printing, binding, stamp-
ing, embossing, and other branches of book and newspaper work;
there are machines, apparatus, and tools for lithography, color printing, photo-mechanical and other mechanical processes of illustrating; for working metals, minerals, and woods. Finally there is a
collection of fire engines and fire extinguishing appliances, whether
by water or chemical apparatus… .

There were elevators and moving walkways. There were "overhead
highways" of various designs. The buildings were electrically lit and outfit-
ted. Besides the main exhibition venue where these massive, awesome
buildings were located, there were supplementary exhibits of a more enter-
taining and less rigorously authentic sort outside the gate, and conferences
and events lasting for a single day or longer organized in conjunction with
it. Some were of already prestigious fields, such as the International
Mathematical Congress, attended by Felix Klein and many other European
mathematicians. It was the occasion of some firsts, though, such as the
U.S.'s first International Aeronautical Congress, held as one such supple-
mental event on August 1–3, 1893. It was organized by Alfred P. Zahm,
with significant involvement and advice from Octave Chanute. Though
Chanute was worried about the reception of a Congress dedicated to a
topic often lampooned and still considered unscientific by some scientists,
it turned out to be a great success, both in the quality of the program and
the positive press it received. It attracted international visitors, notably the
otherwise-isolated Australian Lawrence Hargrave, who had invented the
box kite, had done experiments to determine the best shape for inclined
surfaces for gliders, and had constructed and successfully flown several flying machines. Chanute was happily surprised at the interest and respect
the aeronautical convention received.

Wilbur and Orville Wright attended the fair, but the biographer Tom
Crouch reports that all that is known of their thoughts on it is that they
remarked on the bicycle exhibits. Given that Wilbur Wright said his interest in flight had never ceased since his childhood, and given his nephews'
reports that he and Orville continued to build larger versions of the
Pénaud-style helicopter toy, this would be surprising but for several things.
First, the exhibition was vast, the list of exhibitors and events extensive,

and it is easy to imagine that the announcement of a three-day event in August might have gone unnoticed, particularly by someone who visited in the spring. Besides, Wilbur and Orville had not yet begun their study of aeronautical literature in 1893, and they had neither met nor corresponded with Chanute or anyone else interested in aeronautics and flying machines. The two were in business together in their print shop (building printing presses as well as printing materials to order) and were just starting their bicycle business. And, the exhibits on bicycles, though a very tiny portion of the 1893 World's Fair, were extensive. Bicycles were not humdrum appliances at that time. Making a bicycle controllable enough to be safe and practical had been a major technological feat accomplished only in 1887, and further technological advances were constantly being devised to make them lighter, faster, and more robust. Bicycles had only quite recently gone from being unsafe, unwieldy, impractical curiosities to safe, practical, efficient means of transport (hence the name "safety" used then for the bicycle style we know today, as opposed to the name "ordinary" for the earlier models with a large front wheel).

Crouch reports that the bicycle manufacturing industry "enjoyed phenomenal growth. The number of manufacturers in the field climbed from 27 to 312 in only seven years; total production, estimated at 40,000 machines a year in 1890, reached a peak of 1.2 million by 1895." The reason, he says, is that "[t]he sheer exhilaration of cycling captivated a generation of Americans ... Nothing in their experience could compare with the thrill of racing down a steep hill into the wind, and the newfound sense of personal independence was irresistible." Serious bicycle races became a competitive sport.

When they visited the Columbian Exposition in 1893, Wilbur and Orville had just opened a bicycle business in addition to their print shop. It started with bicycle repair, but they also sold bicycles, parts, and accessories. Crouch says this second business "was literally thrust upon them" for "[t]he two young men who had constructed printing presses from scratch were already legendary mechanics on the West Side. Now they found themselves besieged by friends in need of bicycle repairs." When I hear the Wright Brothers referred to as "bicycle mechanics," it always seems unfair to me that this particular phrase has somehow stuck without understanding its import at the time. It is one thing to train as a mechanic of some fairly well-understood machine that comes with a manufacturer's maintenance and repair manual based on extensive experience with such a machine. It's quite another to be a mechanical genius who can diagnose and fix a recent invention for which there are a multitude of different designs, little maintenance experience with any of them, and no trained mechanics to learn from. The Wrights' mechanical skills were not limited to bicycles; they had shown an ability not only to repair, but to build from scratch, a variety of machines.

Orville, according to Crouch, "seemed destined for commerce" from his school days. Perhaps that explains why someone whose father was a bishop, whose mother had attended college, and whose sister would graduate from Oberlin and teach Latin might be happy in such an avocation. But the fact that Wilbur ended up running such a business for a living requires some other explanation, for he hated commerce. Wilbur had been an extraordinarily gifted student and distinguished himself athletically. When his family moved to Dayton, Ohio in 1884, he took courses to better prepare himself for his expected entrance into Yale College. The explanation his father gave for why he did not become a teacher of some sort, as he had planned, is simple: everything changed when Wilbur suffered an accident "while playing a game on skates." He felt like something of an invalid for many years after. As a result, he did not go to college, but, according to his father's diary, instead he "for a few years, pursued a large course in reading, which a retentive memory enabled him to store for future use."

Crouch reports that Wilbur and Orville were not particularly close before 1889 (again, the year Wittgenstein was born!), "the year in which Orville left high school and Wilbur emerged from a period of extended illness and depression." That depression began with the accident:

> Wilbur had always assumed that he would attend college. "Intellectual effort is a pleasure to me," he told his father, [and later wrote that] "I have always thought that I would like to be a teacher. [...] It would be congenial to my tastes, and I think with proper training I could be reasonably successful." But that dream was now beyond his grasp. [...] Unable to chart a new course that appealed to him, Wilbur fell into a depression born of frustration, indecision, and self-doubt.

He stayed at home, took care of his mother, who was dying of tuberculosis, and read and studied. Though he despaired of getting the college degree he knew he would need to become a teacher, he took full advantage of the intellectual opportunities available to him in his educated and well-read parents' library. Both Wilbur and Orville lived at home with their parents, which, according to Crouch, was not an uncommon occurrence for people of their ages at that time. "All across America young people were finding it much more difficult to strike out on their own than their parents and grandparents had. Times were hard." It was Orville who was especially interested in the printing business, and Orville who started it, but it was with Wilbur that he "designed and built his first professional press out of a damaged tombstone, buggy parts, scrap metal, and odd items scrounged from local junkyards" so that he could expand his business to larger jobs. When Orville wanted an even bigger printing press, again, in 1888, Wilbur helped design it. Orville, ever the businessman, saw the opportunity for profit in a newspaper and began publishing one in March 1889. A few months later, their mother Susan died.

Though Wilbur had been only an assistant in Orville's printing business until then, the newspaper gave him an outlet for the sort of work for which he was especially well prepared. In addition to printing local news not available in the big-city newspapers, they subscribed to a wire service and so Wilbur could read, digest, and choose headlines for national and international stories on a wide range of subjects, from science and technology to international politics. Wilbur began to write an editorial column as well. But their newspaper venture could not compete in an overcrowded market with rapidly improving technology, and they were soon left with just the printing business and the local renown they had earned for building a remarkably efficient press of their own design about whose principles of operation there was some mystique. Again, Wilbur found another outlet: manufacturing printing presses of his own design. He merged this manufacturing venture with the bicycle shop. Thus it was that, by the time the World's Fair of 1893 came to Chicago, Wilbur found himself a close business associate of his younger brother, in the bicycle business that was "thrust upon them."

One commentator has described the significance of the 1893 exposition as follows:

> The Fair's official ideology was an attempt, in large part, to assert a
> sense of American unity as a bulwark against the fear of change
> through pride in the country's accomplishments. It was asserted, in
> the Exposition's architecture, that America had reached cultural par-
> ity with Europe, through its appropriation of the European Beaux-
> Arts form, and through its emphasis on education throughout the
> Fairgrounds. The two areas in which America was already considered
> an international leader, commerce and technology, were celebrated
> extensively in the thousands of exhibits and the placement of the
> Electricity Building and the Manufactures and Liberal Arts building
> directly on the Grand Basin, counterparts to the former bedrocks of
> American society, the Agriculture and Machinery Buildings.

The next year, Wilbur once again considered college. One has to wonder whether the exposition stirred his interest in education or in getting to
a different sort of place in the world than running a manufacturing business in Dayton, Ohio. But for whatever reason—his age, lack of money, the
spectre of financial dependence upon an aging parent and abandoning
responsibility to a younger brother, or some other reason—he dropped the
idea. A year or so later, in 1895, he and Orville decided to extend the bicycle shop into the business of designing and manufacturing their own bicycles, using what they had learned about the strengths and weaknesses of
the bicycles they had repaired and sold. Setting up their own machine shop
gave them opportunities to design some of their own equipment: an internal combustion engine to power the machine shop, an electrical welding
apparatus. As for the bicycle design, they "designed their own oil-retaining
wheel hub and coaster brake. They had no intention of mass-producing
bicycles after the fashion of the large manufacturers. Each of their
machines was a hand-built original made to order." Their business was a
success. They were finally well established in the world. Crouch remarks:
"Most of their friends and neighbors on the West Side must have assumed
that the Wright boys would be pleased to spend the rest of their lives split-
ting their time between the print shop and the bicycle business. In fact,
their attention had begun to wander."

It is surprising there is not more evidence of interest in flying machines in their activities at this point in time. Surely Wilbur must have read about Langley's successful model experiments of powered "aerodromes" and of Lilienthal's gliding experiments, for he was involved in the news business and he had a keen interest in technology. We know from memoirs by people who knew him then that the interest in flying machines on the part of both Orville and Wilbur was early, intense, and continuous. Orville's teacher recalls his saying that he was "assembling the parts of a flying machine, a larger version of which might enable him to fly with his brother." Their nephews' remembrances of them mention that they were building helicopters of various sorts into adulthood, a further indication that the interest was sustained. For whatever reason, though, the kind of methodical self-education and experimentation they had devoted to designing and building their own versions of two of the most significant inventions of their age, one in communication and one in transportation (printing presses and bicycles), were not directed to flying machines until 1896. Crouch points out that numerous people have remarked on the similarity between bicycles and flying machines: "[James Howard Means] ... urged those who sought to fly to pay serious attention to the bicycle."

Crouch calls 1896 "the year of flying machines," citing three events: Langley's successful aerodrome experiments in May of 1896, the successful flights of a glider designed by Chanute and Herring that was flown by Herring on June 22, 1896, and Lilienthal's successful demonstration of his gliding experiments in Germany to an American reporter in August 1896 (followed by his death in a glider crash a week later). Wilbur himself said that his interest in flight was rekindled upon hearing of Lilienthal's death. Perhaps it highlighted to him that here was a problem that even educated people with lots of experience were not able to solve, and hence something that was an opportunity for him. However, almost three years passed before he wrote to the Smithsonian Institute, headed by Samuel Langley, for information about research into flying machines.

This was something new for Wilbur, something other than using his wits in a commercial venture. The printing presses and the bicycles he had designed did employ his ingenuity, but the design of a flying machine was different. The basic problem, the problem of control, had not been solved—except by nature, as in the flight of birds. Just as he had been confident in his abilities to design a bicycle superior to everyone else's, he felt confident that he had as good a chance to solve the problem of flight as anyone else. Perhaps he finally felt he had found something that was his own, something he was meant to do. Certainly there was an intellectual aspect to his analysis of what was wrong with the way others had approached the problem that must have been satisfying. Yet it was an analysis that drew on his practical physical experience with bicycles as well as his grasp of principles. He saw that everyone else was following Pénaud's lead in that they sought aircraft designs that were inherently stable. Bicycles were not inherently stable, but the whole nation was whizzing around on them, going where they wanted when they wanted. As Crouch observes:

> Wilbur's philosophy of control was diametrically opposed to Langley's. He believed that the operator of any vehicle ought to have a means of controlling the motion of his craft in every available axis—an idea firmly rooted in his experience as a cyclist.

> The bicycle differs from all other surface vehicles in that it is inherently unstable in both yaw and roll. The cyclist must steer with the handlebars while at the same time maintaining lateral balance through subtle shifts in body position that will keep the machine upright.

> [... Wilbur] realized that one could take advantage of the subtle links between control in roll and yaw to produce a more manageable, and therefore safer, aircraft.

How did the bird do it? Here the answer calls to mind Boltzmann's remark about rotational means of mechanisms. In Wilbur Wright's own words: "The thought came to me that possibly it adjusted the tips of its wings ... turning itself into an animated windmill." These words bring to mind the image of a large bird turning one wingtip up and the other down while soaring in the sky. That imagery is striking, for, as Boltzmann had noted, Lilienthal had used wing-flapping, which was a reciprocal motion, not a rotating motion. Wilbur Wright observed exactly the same phenomenon as Lilienthal had—soaring flight—but he picked out a different aspect of what birds did in soaring flight as the action that enabled them to fly. Boltzmann had predicted that a rotating device would propel the flying machine that would eventually succeed, but he had thought of rotation as something a bird does *not* do, as something seldom found in nature, in fact. There is something lyrical about Wilbur seeing a "windmill" in the bird's flight. He saw the rotational motion in the means of the bird's control. The bird maintained control of one kind (yaw) by the windmill-like motion of another (roll): moving its wingtips, which tilted one wing surface toward the front of the bird and the other toward the back of the bird. This is like a windmill; windmill vanes are inclined. Imagine a windmill with just two vanes: as they turn, one vane is inclined toward the sky, and the other toward the earth. The soaring bird's wings are not like fixed vanes, however; the bird can move its wingtips and so can tilt a wing first one way and then the other. An "animated windmill."

Next came the insight about how a flying machine might do the same: the moment when Wilbur, pondering these ideas, twisted the cardboard box that an inner tube had come in and saw how to "animate" a box kite in the same way. While many schemes for steering planes were limited by the wind, Wilbur and Orville's scheme *used* the wind, as a windmill did, and as they now realized birds do. The construction of a small model to validate the general idea of wing-warping was the next step, followed by the building of a full-scale model to test unmanned: to fly as a kite. With strings attached to the ends of the glider wings, the kite could be "animated" in the same way. The basic idea was conceived in a matter of months; they made great confident leaps in the months after Wilbur had written to the Smithsonian for information on work done to date by others. When Orville Wright was asked, years later, in an interview, what the most exciting part of inventing the airplane had been, he answered that it had occurred even before he had ever flown, as he had lain in bed, realizing it was possible to fly.

There were many practical details to attend to—selecting and locating appropriate materials from which to build the plane, and so on. Perhaps their business experience in bicycle design and manufacturing had honed their skills in seeing to such practical details, for they proceeded quickly in building the full-size kite on which to test the wing-warping method modeled on a bird's "turning itself into an animated windmill." Flying depended on having enough lift, and that depended on the relative speed of the machine with respect to the surrounding air. Maxim got speed from his steam engine, but, for an unpowered kite, a strong wind would be needed. They needed to find a sufficiently windy location to proceed with their experiments.

Wilbur proceeded methodically again, writing to find out weather information at potential gliding sites. These next tests would require a trip and substantial time away from their business, so they had to wait to test it out until their bicycle business permitted them a vacation. As for sufficiently windy locations, Wilbur's letters of inquiry were met with a warm welcome from the postmaster at Kitty Hawk, North Carolina. It was, as the bird flies, not too far from Richmond, Virginia, where the Wrights had lived when Wilbur was in high school. But practically speaking, it was a world away. It was remote, and so sparsely inhabited that the post office was the postmaster's house and the only building there. They had to plan carefully, for they would have to pack what they needed or buy it en route. In October of 1900, they arrived at Kitty Hawk and assembled the glider.

They mostly flew their new construction as a kite, but when conditions were right, they actually lay down and flew on the glider. Incredible. They were flying in a new way, using principles of control nobody else had thought of—or so they thought. Actually, they were not the first to notice how birds used their wingtips. But their other insights about control were uniquely theirs. Wilbur was not eager to share his design, but he did write two articles on his insights so far; neither gave too much away. One ("The angle of incidence") was published in Britain in the Aeronautical Society's journal; the other ("The horizontal position during gliding flight") was published in Germany. The Wrights had become part of an international community of aeronautical researchers.

There were more improvements they wanted to make, and they constructed a bigger version, with some changes. In designing the 1901 glider, they used information from Lilienthal's table, as presented by Chanute. They flew this glider at Kill Devil Hills, a few miles south of Kitty Hawk. That's when they hit a wall. It didn't work the way they had expected it would, and they didn't see any reason why. For awhile, Wilbur was too frustrated to continue.

Wilbur finally did overcome his frustration, though, after reasoning things through. The only thing that could be wrong, he concluded, was the experimental data he had used from others. So he decided to generate his own data. It was urgent; there was no time to build the kind of wind tunnel that the large institutions had. A box and fan would do, if you were clever enough about what it was you needed to find out. What they really wanted to find out had been calculated from tables of data collected by others, and in terms of one coefficient piled on another, yielding a coefficient for comparing the forces on a certain shape of wing as compared to a flat plate. Why not *directly* compare what the wind does to a model wing shaped in a certain way with what the wind does to a flat plate? This they did by clever mechanisms, using a bicycle wheel and a balance located inside a cardboard box. They carried out a tediously thorough set of experiments, keeping careful records. What they were really measuring was the relation between two things—the relation of the force on a curved or irregularly shaped surface to the force on a flat square plate. Though seldom mentioned, their reasoning involved inferring from the relation between the model wings to the relation between a similarly shaped set of larger wings. It was really reasoning from quite a high level of abstraction.

Wilbur was rewarded: the experiments indicated that the calculations had given them incorrect results! At first he concluded that Lilienthal's tables had been wrong. In retrospect, it is now realized that the instructions he had for how to use the information in Lilienthal's tables was unclear regarding all the coefficients (especially the coefficient associated with aspect ratio) that needed to be applied. In fact, Lilienthal's experimental data had *not* been wrong, and Wilbur would later come to see that on his own. That would all be realized much later; what mattered in 1901 was that Wilbur produced the information he needed, from which he could figure out how to build the next glider. More trials, more modifications, all done in conversation with his brother Orville.

Success followed; they found they could control where the glider went. Next came building a light engine to power the flyer. Again, the demands of the bicycle business intruded, and they had to wait until December of 1903 until they could take the flyer to Kitty Hawk for trials. On December 17, 1903, they got their confirmation: they had solved the problem of powered, controlled, human-carrying flying machines. They sent a telegram home to their sister and father, telling them to inform the press of the facts of their successful flight.

It might have been expected that the news that this age-old problem had at last been solved would have shot around the world at the speed of electricity. But it did not. For one thing, the Wrights were not solicitous of fame. They wanted credit, but that is not the same thing as publicity. In fact, Wilbur especially worried that fame would encourage snoops and spies and imitators and rob him of credit. Hence, the Wrights did not enter competitions nor send around photographs of their flyer in the air. Chanute was less cautious. He did not share Wilbur's concern, though he did try to give the appearance of taking it seriously for their sakes. He thought the solution to Wilbur's worry was simple: file for a patent immediately, and don't worry about it. Chanute had plenty of patents himself. Not only was Chanute's view naive for an invention of such significance as solving the problem of heavier-than-air flight, as future events would show, but Chanute did not really understand that Wilbur did not see himself as continuing a group effort, with his contribution just happening to be the last straw that tipped the scales and resulted in a successful flying machine. Rather, Wilbur thought he had seen where everyone else had gone wrong. Wilbur also thought he had had a number of lucky breaks that had made all the difference in how rapidly their work had progressed, and so that it was unlikely that anyone else could have built a flying machine by the time he and Orville had. But the overwhelming reason why the news did not instantly spread around the world was skepticism. There was good reason to be skeptical, too, for there were many fraudulent claims about flying machines. In fact, some of the news reports about the Wrights exaggerated facts about their flights. For many, the only acceptable proof was participation in public exhibitions.

The 1904 World's Fair in St. Louis was an opportunity for such a public exhibition. There was great interest in flying machines by this time, and Europeans eager to find out what was new crossed the Atlantic to attend. Chanute exhibited a modified and improved version of the glider he had flown in 1896 at Dune Park in Indiana. The Wrights were invited to exhibit the flying machine about which so much had been claimed, but after inspecting the field on which they would have to take off and land and determining it was too rough, they declined. This did nothing to instill public confidence in their claims.

There was one significant exception. Colonel John Edward Capper, commander of the British Army's Balloon Section (where Samuel F. Cody was employed), was in America to visit the aeronautical exhibitions at the 1904 World's Fair in St. Louis. Though skeptical himself, he had heard of the claims made about the Wright Brothers' flyer from Chanute and from a British colleague who had visited the Wrights in 1902. He made the trip to Dayton, leaving as a skeptic and returning as a believer. In his recent book, *To Conquer the Air: The Wright Brothers and the Great Race for Flight*, James Tobin reports:

> Before leaving St. Louis, the Englishman spoke with a reporter about the Wrights and expressed "grave doubts concerning the veracity of even the comparatively modest results they were claiming." Yet only a few days later, the same reporter ran into the colonel in New York and found him "positively enthusiastic" about the Ohioans. Capper had not seen the machine fly. Of the machine itself he had seen only the engine. He had done nothing more than share breakfast with Wilbur Wright and look at some photographs. Yet Capper came away entirely confident that the brothers had done what they said.

As things would turn out, this would not be as significant to the transfer of heavier-than-air technology to England as it first appeared. Capper was no fairer to the Wrights than he was to Samuel F. Cody. He had his own favorite investigator he wanted to fund to develop the aeroplane. To get as much information as he could for his own use, he had led the Wrights to think he was interested in their plane for use by the British Army. Then he laid down impossible conditions for them to meet in order to gain a contract. He had been just as untrustworthy with Cody, whom he got to develop a lightweight engine for the Army—by leading him to think that the engine he was contracted to design and build was for an aeroplane for which Cody would get some recognition, when (as Cody found out only upon finishing the work) the engine he developed was instead used for powering a balloon (the *Nulli Secundus*), the very technology Cody had derided. Capper instead asked an Englishman, Dunne, to develop an aeroplane. Capper's visit to the Wrights on his 1904 visit to the St. Louis World's Fair did do something significant for the development of heavier-than-air flight in Britain, though: it convinced the skeptical British that Americans were ahead of them in developing a practical aeroplane.

There may have been another reason why Wilbur chose not to exhibit the flyer in 1904. He was still not satisfied with the control mechanisms, and he and Orville were still perfecting the flyer. My opinion here is based on the encyclopedia entry that Orville wrote for Wilbur; he says: "Experiments were continued in 1904, but it was not until Sept. 1905 that they [referring to himself and Wilbur] learned to avoid the 'tail-spin' in making short turns." They continued to perfect it; they were finally satisfied when, on October 5, 1905, "Wilbur Wright flew for 38 min. over a small circular course covering a distance of 24 miles. Believing the machine now to be developed to a stage of practical usefulness, the Wrights spent several years in finding a market for the invention." Wilbur had a vision of what his invention was capable of when fully realized, and he knew it was not yet perfected in 1904. In 1905, he would announce to the European aeronautical clubs and journals what he had accomplished, and he would invite them to verify what he and Orville had done by contacting the witnesses he named. But in 1904, he was not quite at that point, and the 1904 World's Fair did not include the Wright flying machine.

The 1904 World's Fair in St. Louis brought Ludwig Boltzmann to America, but not mainly from his curiosity about flying machines. Rather, one of the Congresses arranged in coordination with the exposition was a set of lectures in the humanities and sciences, and Boltzmann was invited as one of its speakers. There had been lectures and conferences at other expositions, but this one was planned to be a much more unified affair. Organized by the German-born Harvard professor of psychology Hugo Munsterberg, it was organized around the theme "Unity of Knowledge." A small committee traveled to Europe in 1903 to promote and explain the purpose of the Congress and to personally invite prominent scholars, offering generous funding for their visits. It aimed to be, and was, an international affair. It brought Wilhelm Ostwald to St. Louis, as well as Boltzmann (who was then in Vienna); Paul Langevin and Henri Poincare from France; Ernest Rutherford, speaking on radiation; Simon Newcomb (whose attempts to ridicule manned heavier-than-air flight were well known); and Edward Leamington Nichols, the German-educated editor and cofounder of *The Physical Review* and a physicist at Cornell University.

Ostwald was asked to give the lecture on the methodology of science, whereas Boltzmann was placed in the section on applied mathematics. In

his introduction, Albert Moyer notes that "Ostwald stressed the importance of establishing a definite and complete 'correspondence' between the
actual 'manifold' of the scientists' conceptual constructions whether they
be words, equations, or symbols." He also reports that, according to Robert
Millikan, the "question of whether the atomic and kinetic theories were
essential" was the Congress's "chief subject of debate." Though viewpoints
on this particular question varied widely, there was much agreement that
there were new phenomena in physics that did not fit into the current conceptions of mechanics. Moyer describes Edward Nichols, editor-in-chief of
the *Physical Review*, as a "more moderate" member of "the progressive
members of the American delegation" and as having "defended the
increasingly familiar idea of evaluating scientific concepts through use of
the 'dimensional formula.'"

The meeting was a major event, and it gives a snapshot of science (here
science is to be taken in its broadest sense, as any field of inquiry) just
before it was to change dramatically; Einstein's suite of papers in 1905 on
relativity, Brownian motion, and the photoelectric effect (light quanta)
would appear the next year. Many of the speakers spoke on the existence of
contradictions. Poincare discussed the contradictions in physics that arise
in attempting to preserve the principle of relativity along with certain facts
about the velocity of light. Rutherford talked about radioactivity. Langevin
spoke of the vast consequences of "the new idea, based on the experimental fact of the discontinuous corpuscular structure of electrical charges."
This included the fact that "they throw a new light even on the fundamental ideas of the Newtonian mechanics, and have revived the old atomistic
ideas and caused them to be lifted from the rank of hypothesis to that of
principles." Boltzmann spoke about the new field of "statistical mechanics," which he used to describe the science systematized by Gibbs: "a new
science ... whose problem is, not the study of the motion of a single
mechanical system, but of the properties of complexes of very many
mechanical systems which begin with a great variety of initial conditions."
This new science arose from a contradiction. Boltzmann did not claim that
the contradiction as stated had been resolved, exactly, but that we "are
cured of the effort to answer it in a senseless and hopeless manner." This
sentiment, that the value of a new foundation of physics might lie in its
therapeutic value of "curing" people of trying to give certain kinds of

answers to certain questions, echoes a line from Hertz's *Principles of Mechanics* that Wittgenstein was "fond of quoting."

The contradiction Boltzmann was referring to was related to the burning topic of the day—the question of atomism. That contradiction concerned the composition of matter: is it continuous, or is it composed of atoms? He said then:

> It appears now, that we are unable to define the infinite in any other way except as the limit of continually increasing magnitudes, at least no one has hitherto been able to set up any other intelligible conception of the infinite. Should we desire a verbal picture of the continuum, we must first think of a large finite number of particles which are endowed with certain properties and study the totality of these particles. Certain properties of this totality may approach a definite limit as the number of particles is increased, and their size decreased. It can be asserted, concerning these properties, that they belong to the continuum, and it is my opinion that this is the only self-consistent definition of a continuum which is endowed with certain properties.

But what he calls here a "self-consistent definition" is consistent in the sense that the "old philosophical question" no longer bothers us, not in the sense that it has actually been answered in the terms in which it has been stated. The question has been transformed:

> The question if matter is composed of atoms or is continuous becomes then the question if the observed properties are accurately satisfied by the assumption of an exceedingly great number of such particles or, by increasing number, their limit. We have not indeed answered the old philosophical question, but we are cured of the effort to answer it in a senseless and hopeless manner. The thought-process required—that we must investigate the properties of a finite totality and then let the number of members of this totality increase greatly—remains the same in both cases. It is nothing other than the abbreviated expression in algebraic symbols of exactly the same thought when, as often happens, differential equations are made the basis of a mathematical-physical theory.

What he is referring to here is the basis of the symbolic system used in mathematics for so many physical processes: the differential calculus. When first formulated, the arguments for the truth of the fundamental theorem of calculus given by Newton appealed to geometric intuition. The explanations he gave—and Newton kept reformulating the calculus to provide more defensible accounts each time—depended on the notion of a limit (the value of a ratio as the numerator and denominator were made to decrease in size indefinitely or, alternatively, were made infinitely small), and some questions were raised about determining limits. Attempts to establish foundations for the calculus had caused a crisis in other parts of mathematics. Gottlob Frege's work in logic and the foundations of arithmetic was meant to put mathematics on a logical foundation; to provide a formal language in which it could be expressed, so that it would be clear which thoughts depended on which others, and hence so that which proposed proofs of the foundations of the calculus were logically valid and which were not could be objectively determined without appeal to geometric intuition. Hilbert's addresses in 1900 put forth the questions that mathematicians should answer in order to put mathematics on a sound logical basis.

There was more to Boltzmann's lecture than these points about mathematical formalism; he went on to point out that many questions still remained, even after the question of atomism had been dealt with in this manner. For even if continuous quantities can be dealt with in this manner conceptually, and thus questions about continua turned into questions about mechanics, the laws of mechanics do not prefer processes that go from more highly ordered to less-ordered states. So it seems the second law of thermodynamics does not follow directly from the laws of mechanics. Boltzmann discussed the assumptions that would need to be invoked to show the second law of thermodynamics a consequence of mechanics.

These latter points are generally considered more central to the problems of statistical mechanics, as debates in science and physics journals, to which both the then-unknown Albert Einstein and Edgar Buckingham contributed, make abundantly clear. The philosophical issues arising between Ostwald and Boltzmann are especially significant to our story, though: the thought processes that are involved in physics employing the symbolism of differential equations, and Boltzmann's tactic of reformulating a question so that "we are cured of the effort to answer it in a senseless and hopeless manner" rather than dealing head-on with a "philosophical question."

The Congress in 1904 was a major event, and Boltzmann's reply to Ostwald in their historic meeting in America was surely of great interest in Vienna as well. Moyer cites a contemporary account of the Congress that reported that Ludwig Boltzmann was seen "sitting in the front row for Ostwald's lecture and joining in the following discussion" and referred to him as "Ostwald's scientific adversary from Leipzig and, more recently, Vienna." He also describes Boltzmann's talk the next day as "a formal rebuttal." Whether or not Boltzmann intended it as a rebuttal, this account indicates that their talks were perceived by the audience as continuing a debate begun on the other side of the Atlantic Ocean.

The significance for our story is that this historic international meeting of major physicists in the year just preceding Einstein's *annus mirabulus* indicates that questions were brewing about just what role symbols and equations play in physics, with Ostwald giving special place to "words, equations, or symbols" and Boltzmann giving special place to "thought-processes." Boltzmann's notion covers a wider range than "words, equations, or symbols," as it includes conceptualizing matter as composed of atoms, and imagining the number of atoms to be very large—even infinite.

Nichols' talk, on "dimensional equations," brought up a different kind of equation than the differential equations Boltzmann spoke of. In a way, dimensional equations are a more abstract kind of formulation than scientific equations, yet the symbols themselves are more constrained by physical theory than they are in scientific equations. In equations used in science, whether differential or algebraic, the choice of symbols is rather unrestricted, and physical relationships are expressed by the form of the equation. In dimensional equations, symbols work somewhat differently. Nichols' characterization is not at all general, and would not be accepted today, but his paper gives us a snapshot of the understanding of dimensional equations by at least some physicists in 1904:

> The science of physics ... has for its foundations three fundamental conceptions: those of mass, distance, and time, in terms of which all physical quantities may be expressed.

Physics, in so far as it is an exact science, deals with the relations of these so-called physical quantities; and this is true not merely of those portions of the science which are usually included under the head of physics, but also of that broader realm which consists of the entire group of the physical sciences, ...

The manner in which the three fundamental quantities L, M, and T (length, mass, and time) enter, in the case of a physical quantity, is given by its dimensional formula.

Thus the dimensional formula for an acceleration is LT^{-2} which expresses the fact that an acceleration is a velocity (a length divided by a time) divided by a time. Energy has for its dimensional formula L^2MT^{-2}; it is a force, $LT^{-2}M$ (an acceleration multiplied by a mass) multiplied by a distance.

What Nichols has done is to specify up front how many quantities there are in physics and what they are. This is not at all a necessary part of working with dimensional equations. In fact, his paper almost begs the reader to ask why the formalism of dimensional equations cannot be extended to deal with the new kinds of quantities in physics indicated by the new phenomena such as alpha, beta, and gamma rays and electrical charge, not to mention more familiar phenomenological quantities such as temperature not included in Nichols' three fundamental quantities. His view precludes this, however, for he states his conviction that "all physical quantities may be expressed" in terms of the "three fundamental conceptions" of mass, distance, and time, as a criterion for assessing a theory:

Not all physical quantities, in the present state of our knowledge, can be assigned a definite dimensional formula, and this indicates that not all of physics has as yet been reduced to a clearly established mechanical basis. The dimensional formula thus affords a valuable criterion of the extent and boundaries of our strictly definite knowledge of physics.

His tone is that of a Cartesian investigator, groping for the things he can be sure of in the face of a challenge to what can be known:

Within these boundaries we are on safe and easy ground, and are dealing, independent of all speculation, with the relations between precisely defined quantities. These relations are mathematical, and the entire superstructure is erected upon the three fundamental quantities, L, M, and T, and certain definitions; just as geometry arises from its axioms and definitions.

As for the quantities used in making "electric and magnetic measurements," he says that the need to add quantities such as magnetic permeability or electrical charge indicates "our ignorance of the mechanics involved."

It is quite odd that Nichols would write this; one of the earliest uses of dimensional formula in physics was by Fourier, in his work *The'orie analytique de la chaleur* (*The Analytic Theory of Heat*). Fourier put no such restrictions on the quantities to be used in physics; he wrote:

> It should be noted that each physical quantity, known or unknown, possesses a dimension proper to itself and that the terms in an equation cannot be compared one with another unless they possess the same *dimensional exponent*.

> In the analytical theory of heat, each equation represents a relation between coexistent quantities such as, length x, time t, temperature v, heat capacity per unit volume c, surface conductivity h, and thermal conductivity k. Such a relation does not depend on the choice of the unit of length which, by its very nature is arbitrary.

Nichols' use of dimensional formula thus seems forced, as though it is brought in to justify a criterion of physical theory not at all implied by it. He discusses dimensional formulae only in conjunction with the conviction that physics has exactly three fundamental conceptions. That he employed dimensional formula at all, however, indicates the growing interest in such formalism. As we shall see, dimensional formulae were also becoming important in engineering. The crossover from physics to engineering was due in large part to Osborne Reynolds, a British scientist and engineer at Owens College in Manchester, England. He had employed dimensional equations in scientific investigations, but he was also the first full-time engineering professor there and had developed the Manchester engineering facilities from very modest ones into an extensive laboratory with scale models of canals and rivers, to be used in hydraulic engineering experimental investigations. We shall look at some specific and very remarkable papers by Reynolds using this sort of reasoning later.

For now, we are following the path of Wittgenstein from childhood to youth. The year after the St. Louis Congress, 1905, Boltzmann's *Popular Writings* was published. The collection included Boltzmann's 1894 lecture "On Aeronautics," as well as the paper delivered at the St. Louis Congress in 1904, retitled "On Statistical Mechanics." By this time, Wittgenstein had graduated from the Realschule in Upper Linz, which qualified him only for entrance into a technical degree program, such as his father had attended. The youngest of the family, he had been born into a very financially secure situation, as we have seen, living in his family's grand house in Vienna and the nearby palatial summer home with extensive grounds that his father had bought when Wittgenstein was five years old. The Wittgenstein biographer Brian McGuinness describes the "special atmosphere" in the household:

> The eldest daughter, Hermine ... said that all the children felt it to be one of constant excitement, an unrelaxed tension that came in the first place from their father. His vigour and decisiveness would have put his stamp on any family—a father at the height of his business success ... Karl Wittgenstein was for his children the dispenser of all good things, the creator of the world of large houses, parks and estates which they took as a natural environment.

More generally, he writes that

> They were impressed above all with the importance of honesty, of strict performance of duty, and of fulfillment of obligations towards servants and dependants. Formal religion, in comparison, played little part in their lives. Moral and cultural and material superiority to those that surrounded them (or the consciousness of each of the three) were inextricably interwoven and formed the atmosphere in which they lived.

But their father could be something else, too. McGuinness reports "practical jokes at the expense of those who would feel themselves socially insecure—an entire house-party secretly changing for dinner to embarrass a newly-arrived guest (whom the servants reported to have arrived without evening clothes)." He may have been the dispenser of all good things, but he could also withhold. McGuinness seems to regard it as a kind of inattention rather than a species of cruelty: "Sure of himself and of his values, he imposed them on his sons with not much attention to their gifts and none to their inclinations. They were to learn mathematics and Latin and in later life they were destined—by him—to combine engineering and business as he had done," yet he mentions Poldy Wittgenstein's regret after Karl died that "she said nothing to her husband about his harshness towards Hans." Though the children were tutored at home, nobody ensured that the tutors were capable or effective. Thus, Ludwig's older brother, Hans, came to an unfortunate end:

> ... he had no interest in the career chosen for him and he had not learnt such habits of regular even if distasteful application as a normal schooling might have given him. Thus when he was sent, without a proper tutorial training, to pick up the necessary knowledge in various enterprises in Bohemia, Germany, and England, he had no idea what was expected of him... He vanished from a boat in Chesapeake Bay at the age of 26 in circumstances suggesting suicide: as such, certainly, from whatever indications, his death was always regarded in the family.

There had been drama in the household before that, too: in 1898, when Wittgenstein was about nine, his father had retired, in the wake of something like an insider-trading scandal. His father stood by his actions as innocent and their suspicious timing as coincidence. There were sarcastic articles about him in the newspapers. It seems to me that the public "criticism and envy" of the man around whom the entire household revolved and to whom everyone in it answered would have had an effect of some sort on the household, even if it did not dim Karl Wittgenstein's own confidence in the rightness of his actions. One has to wonder about the coincidence of this event and a reminiscence Wittgenstein had of when he "was about 8 or 9":

> When I was about 8 or 9 I had an experience which if not decisive for my future way of life was at any rate characteristic of my nature at that time. How it happened, I do not know: I only see myself standing in a doorway in our house and thinking "Why should one tell the truth if it's to one's advantage to lie?" I could see nothing against it.

Hans' suicide occurred in 1902, several years after his father's "retirement." The next year, the third son, Rudi, poisoned himself in Berlin, where he had been sent to study science. McGuinness tells a heartbreaking story about Rudi's childhood: "When he was seven, on the occasion of a test of elementary education that he had to undergo, he seemed so frightened and miserable that the examiner warned his mother that he was a nervous child, who ought to be watched. This was always quoted by the mother as a great joke." McGuinness notes that Rudi's suicide occurred the same year that Otto Weininger, author of *Sex and Character* and something of a cult figure, also committed suicide. He also grants that although "Ludwig and his brothers were affected by belonging to a society which recognized suicide as an acceptable way out," that is surely not the whole story in their case. He notes Ludwig's "struggle inside himself which led him constantly, for years after 1903, to doubt whether he could ever do anything worth while, and hence to the thought sometimes of suicide, sometimes of impending (because wished for) death."

When Ludwig was fourteen, in 1903, the first decision about his education had to be made. According to McGuinness, that he was sent to the Realschule in Linz rather than to the Gymnasium near Vienna "is perhaps a sign that he appeared more fitted than Paul for the scientific or technical education that such a school would prepare him for." There is the anecdote about his constructing a model sewing machine from available materials. McGuinness speculates that it must have been considered desirable "both to separate the boys and to take them out of the atmosphere of the family home in which they had not achieved either much learning or much happiness" and remarks on the unhappiness and loneliness of Ludwig's childhood. Michael Nedo, too, remarks on a softening of parental attitude, but he notes Ludwig's desire to return to that atmosphere:

> After the suicide of their brother Rudi ..., their father showed more understanding and patience for his two youngest sons. He acceded to Ludwig's wish to stay away from school. His father instructed his wife in a letter that "Lucki ... is to come to Vienna so as to have for the present a chance to laze about properly. If Lucki wants to learn at home, that's fine; if he wants to go into a workshop for the next few months, which he needs to do sometime in any case, that's fine too.... He should laze around all he wants, sleep, eat, let off steam, go to the theatre."

Evidently Ludwig was not very happy at school. He lived with a schoolmaster from another school, and McGuinness reports that one of his classmates remarked on how different he was in almost every way from his classmates, and that "his performance in school subjects was far from distinguished." Significantly, though, "[his brother] Paul thought that his main interest was in physics."

When he left school in the summer of 1906, he wanted to study with Boltzmann. This was within a year after Boltzmann's *Popular Writings* had appeared, an anthology that, as mentioned earlier, included the article "On Aeronautics," as well as Boltzmann's 1904 lecture on statistical mechanics. At this point in time, the Wrights still had not come to Europe to demonstrate their flying machine, and, even though there were reports that they had flown long sustained flights, there was not the sense that they had decisively solved the problem of flight. The aviation enthusiast and promoter Ernest Archdeacon specifically questioned their claims of success, on the grounds that they had not created an inherently stable aircraft.

Even among those who believed there was something to the reports of the success of their flying machine, it was unclear how much of the Wrights' success was due to the design of their machine and how much was due to their (perhaps exceptional) flying skills. Lilienthal had used the method of shifting his weight for control in his gliding experiments near Berlin, and there was no reason to discount the suggestion that the Wrights were displaying an unusual acrobatic ability rather than an unusual invention. Aeronautics was thus still seen as an area in which a young genius might conceivably have an historic impact by solving the age-old problem of flight, since the Wright Brothers' success was still in question in Germany. McGuinness reports that "[Wittgenstein's] sister Mining thought that from the start he was concerned with aeronautics and he certainly came to Manchester in 1908 already full of projects connected with it."

There has been some puzzlement about Wittgenstein's statements that he had planned to study with Boltzmann; McGuinness speculates that "it must have been the philosophy of science in Boltzmann that attracted him," explaining it as "the first signs of his divided vocation." Perhaps, but another possible account attributes more unity to his outlook. Boltzmann had written the article on aeronautics that Wittgenstein could have recalled hearing something about as a child of five when it was delivered in Vienna. And Boltzmann had announced that he would "teach elasticity and hydrodynamics in the winter semester of 1906–1907." Now, these are just the subjects needed for aeronautical engineering. It is true that Boltzmann said experimentation with kites was the way to study the principle of the inclined plane, and that a theoretical field such as hydrodynamics would not be able to deal with the complexities of airflow over a wing. However, if one were going to study an area of physics that is applicable to aeronautics, in order to learn the governing physical principles and equations, that is exactly the subject one would choose. In fact, Hermann von Helmholtz had written a paper on using the differential equations of hydrodynamics to study dirigible balloons or airships, in a different sort of way. Helmholtz had manipulated the form of differential equations that were not soluble by known integration methods to obtain some ratios—pure numbers, with no dimensions—that could be used to say when two situations in hydrodynamics would be similar with respect to motions and velocities. Two situations would be similar regarding their corresponding velocities, for instance, if certain ratios he described were the same in each situation. So the equations of hydrodynamics, though insoluble, could be useful in aeronautics, in a way that related to experimental work. Helmholtz had described how one could use the results of one actual experiment to infer something about experiments not yet carried out, using hydrodynamics. The course on elasticity would not be unrelated to his aeronautical aspirations, either, for the problem of aviation as Boltzmann saw it involved designing the right kind of kite-like apparatus, and elasticity deals with mechanics of materials: bending and strength characteristics. Boltzmann committed suicide in 1906, however, and never delivered those lectures.

Thus Wittgenstein ended up attending the Technische Hochschule (THS) in Charlottenberg, near Berlin, considered "the most renowned and the best of the German engineering schools." McGuinness points out that Wittgenstein's plans to enter the THS were made before Boltzmann's suicide, but it is possible he planned to attend the THS, with a leave to attend Boltzmann's lectures for the Winter 1906–1907 term. Although the THS offered many more theoretical courses than the ones Wittgenstein took, the certificate program in which he was enrolled was a practical one, and it seems to me the kind of thing often done just to get it over with. It can easily be seen as something done in conjunction with studying physics, especially so-called "applied" courses such as hydrodynamics and elasticity. Wittgenstein entered the THS in late October 1906 for a year-and-a-half-long certificate course in mechanical engineering.

According to comments he made later, he felt these months at the THS were a waste of time. However, McGuinness speculates that he drew on some things he learned there in writing the *Tractatus*. Of special interest is McGuinness's suggestion that

> one central idea of that work [the *Tractatus*]—the idea of the proposition as a picture—owes much to reflection on Professor Jolle's subjects, which were Descriptive Geometry (his study of methods of representing solids and other figures in three dimensions by drawings in one plane) and Graphical Statics (a method of reducing a system of forces graphically and showing the resultant force, couple or equilibrium).

Wittgenstein does use both these metaphors—projection, and taking the resultant in a graphical representation—in the *Tractatus*. Whether they suggest the idea of a proposition as a picture is a separate question.

The books Wittgenstein bought during that period give some indication of his interests. Many of the more valuable books he purchased in Berlin from an antiquarian shop have been preserved as an intact collection by a chance event: at one point, Wittgenstein asked that the things he had left in Cambridge be sold, and Russell decided to buy the books. Russell's book collection has been preserved in the Bertrand Russell archives, and along with it at least some of the books he bought from Wittgenstein. These have recently been made the subject of an article (by Oystein Hide); it is striking how many of the books relate directly to aeronautics. The collection includes a six-volume set of da Vinci's work, a rare book by Galileo on mechanics of materials, and some books on engineering design. Another book Wittgenstein very likely purchased in Berlin is a German translation of *Hydrodynamics*, the compendium of the discipline by the Manchester University professor Horace Lamb. This last book was found in a London secondhand bookstore a few years ago by Peter Spelt, then a researcher at Imperial College in London who enjoys browsing in used bookstores. Spelt and McGuinness have coauthored an article describing the marginal notations in the text. How the book got to that store is anybody's guess. But since it was a German translation of a book readily available in its original English, a very plausible suggestion is that it was among the books Wittgenstein left in Cambridge for which he had left instructions to be sold, and was among the books Russell bought, but was one that Russell had no reason to keep.

There is some speculation about what led Wittgenstein to leave Berlin and Vienna for a place as unfamiliar as England, especially considering Germany's strength in engineering. One is that Lamb was at Manchester, and Spelt and McGuinness reinforce that suggestion. Given the relation of hydrodynamics to aeronautics, and Wittgenstein's disappointed plan to study the topic with Boltzmann, this seems quite plausible. Probably England was as good as Germany as far as the next step on the path Boltzmann had laid out: studying the "principle of the inclined plane" by means of a kite. Although Maxim, who was going in the wrong direction, was in England, so was Samuel F. Cody, who was going in the right direction. By this time Cody's unparalleled success with kites was probably well known, for he patented his unique man-lifting kite in 1901. The picture of Wittgenstein with a kite taken in England that appears on the cover of this book does, as I have mentioned, resemble the Cody-Hargrave style of kite. Hargrave developed the box-style kite and determined the curvature the surfaces ought to have; Cody improved upon it. The bat-style wing extensions, which are faintly reflected in the extended corners of the kite pictured in the photograph, were his signature style and his invention.

By 1905, when Wittgenstein entered the Technische Hochschule (THS), Cody had developed and flown a glider-kite of his own design. In 1907, while Wittgenstein was still at the THS, Cody developed a light engine that achieved fame for powering a dirigible balloon, the *Nulli Secundus*, which broke flight records on its maiden flight. Later that year, Cody attached an engine to one of his kites and began making flights in his powered kites, an activity he was still engaged in when Wittgenstein decided to leave Berlin and Vienna behind to go to England. If, as I am suggesting, Wittgenstein's main consideration in choosing where to go next was to design, build, and experiment with kites, England would have been an obvious choice on that account alone, and the other considerations, such as Manchester's Horace Lamb being the best person with whom to study Lamb's *Hydrodynamics* (the standard compendium on the subject), reinforcing considerations. What we do know is that the first place Wittgenstein planned to go in England was to a place where he could build kites, and that, in fact, the first thing he did when he got to England was to learn to design and construct kites. Finding such a place in the world was a step toward a meaningful avocation of his own choice, perhaps the first serious undertaking he had ever chosen for himself.

During his time in Berlin at the Technische Hochschule, it appears that Wittgenstein acquired book after book about the history of flight. Hide's paper listing and discussing these books shows a number of 1784 works on the historical development of aeronautics, many of which were purchased in Berlin from a particular Antiquarian book shop. There is also a copy of Newton's *Principia* (in Latin) and a very rare copy of Galileo Galilei's *Discourses Concerning Two New Sciences* ... , published in London in 1730, the significance to aeronautics of which we shall consider later. So, it appears that Wittgenstein was very interested in the history of flight. His interest may have been as much in the works' description of inventors solving flight problems for the first time as in any technical information they contained. Wittgenstein's time spent at the THS in Charlottenberg had accomplished this much: he had put behind him the requisite year or so in a practical engineering setting that his father expected of him.

A number of commentators think Wittgenstein also read at least parts of Heinrich Hertz's *Principles of Mechanics* by or during this time. There is the fact that he was "fond of quoting from Hertz" the passage in Hertz's introduction: "The whole task of philosophy is to give such a form to our expression that certain disquietudes (or problems) vanish," the sentiment so similar to the one Boltzmann had expressed. Although various commentators on Wittgenstein and Hertz state that he read Hertz's *Principles of Mechanics* even earlier, during his years in Vienna, and consider it a major influence, it is not clear just when he read Hertz. Certainly had he not read Hertz on his own in Vienna or Berlin, he would have learned of Hertz through Russell's work, for the book of Russell's that he said stimulated his interest in philosophy while he was at Manchester (Russell's *Principles of Mathematics*) contained an entire chapter devoted to an analysis of Hertz's *Principles of Mechanics*. The work was well known, and he could have seen or heard it discussed any number of places, including in Boltzmann's textbook on mechanics. Nobody really knows just when or where he first encountered it, due to the myriad of possibilities and the dearth of surviving documents from that time. McGuinness writes "That [Wittgenstein] came to know Hertz before undergraduate days is a conjecture based on his mentioning Hertz first among those that influenced him and also on the fact that Hertz and Maxwell were the sort of authors read in the Allegasse."

On a personal level, Wittgenstein had by 1908 outgrown the role of the polite, charming, and affectionate guest in the homes of the host families who had been arranged for him during his time at school. By the end of his time at the Technische Hochschule, his personal letters indicated a preference for privacy and distance from that kind of closeness with the family with whom he stayed. This is not an unusual change for someone of seventeen or eighteen years of age. For Wittgenstein, it also marked a stage in his development in which he had finally begun finding his own place—in terms of avocation as well as personal space. The place in the world he found for himself—for the time being, at least—was a place where he could learn how to build and fly kites of his own design. Some other things in this place in the world he had chosen for himself were important to him, too: music, philosophy, and foundations of physics. In his later avocation as a philosopher, Wittgenstein would draw on many of these interests from the new place in the world that began with his kite-flying job.

Chapter 4

A New Continent

L udwig Wittgenstein's move to a kite-flying station in the north of England in the summer of 1908 was actually part of a more general move: the move to study aeronautical engineering at the University of Manchester (formerly Owens College). The job at the kite-flying station was a summer position; in return for assisting with "constructing, sending up, and recovering the instrument-bearing kites" used for meteorological observation, he would get to use the equipment there for his own kite research. Apparently he was inexperienced, for he wrote home from Glossop that he first observed and then learned how to make a kite.

Professors from the university ran the experimental station where kites were used for meteorological observation. McGuinness notes that:

> The lecturer in meteorology from 1908 (and thus the person in charge of the experimental station) was a distinguished physicist, J. E. Petavel. Among his other gifts he was particularly ingenious at designing and adapting machinery for experiments and it was for this reason that he was called in to improve the winch used for the instrument-carrying kites and hence became interested in the whole project.

The winch system for that instrument-carrying kite system may well be Cody's man-carrying kite system, and Cody likewise became interested in solving the problem of heavier-than-air flight through such inventions. Wittgenstein had found a place where such inventiveness cohabited in a mind that was interested in the kind of physics he was interested in, too:

Petavel, already a Fellow of the Royal Society and later Director of the National Physical Laboratory . . . had done important work on explosions, devising the Petavel gauge, . . . had worked on low-temperature phenomena, the expansion of gases, gas engines, reactions at high pressures. Now, after a visit to the kite and balloon station at Glossop, he made a free balloon ascent and became interested in aeronautics. He was appointed to the original Advisory Committee set up by the government in 1909 and was soon a leading authority on all aspects of the subject.

At Manchester, there certainly were departmental divisions, and certainly physics and engineering were in separate divisions. But, at Manchester, observes McGuinness, "divisions into departments did not dictate the studies or limit the collaboration of the scientists working there." He notes that Petavel was in physics, and then meteorology, and after that, became Professor of Engineering. The physicist Rutherford, with whom Wittgenstein had expressed interest in studying, was at Manchester by the time he arrived. Being enrolled as an engineering research student would not have meant being separated from the physicists: "The seminars or laboratories or workshops of one department would be made available to researchers from another and members of Rutherford's team would—as a holiday or a hobby—go out to Glossop and try some meteorological experiment with the kites."

Wittgenstein seems to have found a sort of surrogate brother in his work at Glossop: William Eccles, an engineer older and more experienced in engineering than he. According to Eccles' account, when he first met Wittgenstein, Wittgenstein was surrounded by the books and papers he was reading. It seems to have been a relationship somewhat like the relationships that existed between brothers in the many pairs of brothers who have worked together in aviation. "They talked a lot about technical problems, though more those of the work they were jointly engaged in than any arising out of Wittgenstein's work." Then there were the long, trying, physical demands of the joint enterprises they shared, requiring stamina and long stretches of time spent in each other's company: "The work at the station was arduous and continuous. Sometimes there would be eight or ten ascents a day until as late as nine or ten at night. The kites would be sent up as high as 5,000 feet—naturally this demanded a train of kites. Sometimes the kites would escape or come down and then a correspondingly long distance would have to be traversed on foot to recover them. There were the dangers, too, of storms . . ."

The friendship continued for years. Even after he no longer had any formal ties to Manchester, Wittgenstein planned visits there during his vacations to see Eccles and his family. They invented things together and discussed furniture designs together. Sometimes they went together to concerts in Manchester.

Wittgenstein's work in aeronautical engineering followed the lead of those doing research on heavier-than-air flight: he went from kites to combustion chambers, presumably for engine design. These were the days when engine designs became famous by name. The French Gnome engine, in which the array of cylinders itself rotates about a combustion chamber, was well-known and a popular choice, and the mechanics who knew how to work on it commanded respect and high salaries. Wittgenstein studied with Horace Lamb, and, again following the needs of the discipline, he focused on propeller design when that became the national interest. He became interested in the foundations of mathematics. Exactly what, if anything, precipitated the decision not to return to Manchester in 1911 is not known. It is more a matter of too many accounts rather than none; McGuinness lays out a number of different accounts based on the remembrances of those close to Wittgenstein and compares them with the available documents. The story involves giving his sister the impression that he had been suddenly gripped by a strong and irresistible need to study philosophy, it involves a visit to Frege, and it involves Wittgenstein appearing in Cambridge in the fall of 1911 without any advance plans or announcement, expecting to study logic with Bertrand Russell.

Work in Manchester was not without its stresses, but overall Wittgenstein's life there seems to have been relatively happy, perhaps the most content he had been in an environment at that point in his life. Yet eventually, it was with him somewhat like it was with the Wright Brothers once they had established their own bicycle manufacturing business. After Wittgenstein had established himself in a place where he could plan and execute his own experiments, where his ingenuity was not only recognized but actualized—he could and did design and patent his own inventions—there was still a restlessness, a sense that there was something else he ought to be using his talents for, something momentous.

Life in Manchester was comfortable for Wittgenstein in material ways, as well as intellectually and socially. He was well off, having an "enormous" personal income. He could afford to buy the equipment he needed to carry out his experimental work, and he did. He lived, probably again with a wealthy family in Manchester found for him by his own, among "comfortable large houses, with a fair sprinkling of Jews and Armenians. . . . He dressed carefully and apparently expensively." There were more than material amenities, too: the intellectual life already mentioned, and his being a part of a community in which he daily saw the intellectual life of physicists extend to the practical work of employing ingenuity in conjunction with an understanding of physical principles (in their work designing test equipment and solving experimental problems). Manchester was an industrial city that had undergone expansive growth—from around 100,000 in 1801 to over two million (including Greater Manchester) in 1901—and had a sizable German-born population. There was music, too, perhaps in part as a consequence: "every informant," reports McGuinness, "mentions visits to the Halle' concerts for more serious composers— Wagner, Beethoven, Brahms—and the concentration with which he would listen and the enthusiasm with which he would then talk about music."

After Wittgenstein left Berlin and Vienna for England, there was a thunderbolt of a development in the history of flight: 1908 was the *annus mirabilus* of flight as far as European awareness was concerned. If, as numerous commentators report, and as the remembrances of Russell and Wittgenstein's family members support, Wittgenstein's purpose in going to study aeronautical engineering in Manchester was to design, build, and fly an aeroplane, the events of 1908 would have been extremely significant. For it was in the summer of 1908, just after Wittgenstein left the European continent, that Wilbur Wright arrived there for an extended tour. He brought his flying machine with him, the perfected version he had announced in 1905, to a partially supportive, partially doubting European audience. The flying machine, the 1905 Wright Flyer III, was his answer to the problem of practical heavier-than-air flight—a machine capable of sustained, powered, controlled human-carrying flight.

He came, he flew, he stunned people. The reports do not disagree on that. In 1904, Wilbur had had some abortive demonstrations due to poor weather conditions. Not in 1908, though—not with the 1905 Wright Flyer III.

Due to Chanute's dissemination of information about the earlier Wright machines, there were already some imitators in Europe whose machines looked very similar to the Wright Flyer and that were capable of flying short distances. However, these imitators were neither very agile nor capable of staying in the air indefinitely—that is, until their engines ran out of fuel.

"The great impact of the Wrights on Europe," historian of flight Charles Gibbs-Smith has said, "came with the publication of a letter from the Wright brothers to Georges Besancon, in the Paris sporting daily *L'Auto*, in its issue of November 30th, 1905." It was published in *L'Auto* rather than in Besancon's own journal, *L'Aerophile*, which would have been the appropriate place for it, Gibbs-Smith explains, because *L'Aerophile* was running late, and Besancon feared a German journal would beat France to the punch in announcing the important news. The Wrights were spreading the news to the entire aeronautical world. With their new Wright Flyer III, they were beating their previous record with each additional flight, flying indefinitely in perfect circles and returning to the machine's starting point again and again, and landing safely and smoothly each time. The lengths of these beautifully executed flights were limited only by the fuel in the engine's tank. The only thing that had kept the Wrights from setting the time record above an hour (a sort of psychological milestone sometimes mentioned) was that "after flight became more prolonged we were unable to avoid the cars [of curious observers], and the news of what we were doing then spread so rapidly that, in order to prevent the construction of the machine from becoming public, we were compelled suddenly to discontinue experiments." In a postscript to the letter, the Wrights offered to provide the names of "well known citizens of Dayton" who had witnessed the flights described. What effect did this announcement have? Count Henri de La Vaulx describes the general reaction: "there was a general refusal to believe in the veracity of the Wright brothers."

That the Wright Brothers wanted to perform their flights in secrecy and refused to show their machine to reporters no doubt contributed to such a reaction. But when the editor of *L'Auto* sent a journalist to investigate, he reported back, after interviewing the witnesses, that "it is impossible to doubt the success of their experiments." January 1906 prominently featured substantial yet sensational articles on the Wright Brothers' claims, not only in *L'Aerophile*, but in the British *Automotor Journal*.

Still, things were polarized in France. Some thought France should admit it had been beaten, purchase the Wright Flyer, and get aviation going in France with a head start over neighboring nations. Others wanted France to have the honor of inventing flight and so wanted to regard the accomplishments of the Wright Flyer as "tentative." The somewhat disingenuous plan associated with the latter view was to put all the efforts of the French aeronautical community into encouraging development of the aeroplane in France. Sponsoring aerial competitions with extravagant prize purses based on reaching certain milestones in public demonstrations was one way to claim the French had won. The French could be the first, in a way: "If France wishes to do what is necessary, we can still arrive before the others, and present the first demonstration in public of a flying-machine. But we must hurry up." This did have the effect of stimulating aeroplane development in France, and Alberto Santos-Dumont, who had previously used engines to propel cigar-shaped balloons, was able to get a heavier-than-air flying machine off the ground—enough for France to celebrate the achievement of inventing the aeroplane. In January 1908, the Englishman Henry Farman, then in France, was able to meet the more demanding conditions of one of the prizes established: to fly a kilometer from the starting point to a specified target point and return. The entire flight was a minute and a half, a paltry achievement compared to what the Wrights had achieved. More importantly, his control was very crude; it was obvious that Farman did not have the capability of repeatedly going in a circle, as the Wrights claimed they could do. The important thing seemed to be that Santos-Dumont and Farman had made public flights. The Wright Brothers would not agree to public demonstrations until they had signed contracts in hand. They got them in early 1908—one with the U.S. Army, and another with a French company. The contracts required achieving certain feats in public demonstrations.

Finally, the obstacle to seeing the Wright Flyer III firsthand had been removed. There would be public flights. In August 1908, Wilbur made the first public demonstration flights in France. Orville stayed in the U.S. to make the demonstration flights required by the contract with the U.S. Army. In France, Wilbur faced an audience that contained some stubborn unbelievers. In under two minutes, he converted them. The *London Times* of August 14, 1908 printed a report from France:

These experiments were really remarkable. [. . .] They are the public justification of the performances which the American aviators announced in 1904 and 1905, and they give them, conclusively, the first place in the history of flying machines, that rightly belongs to them. [. . .] We beheld the great white bird soar above the race-course, pass over and beyond the trees from its shed to the winning-post of the course. We were able to follow easily each movement of the pilot, note his extraordinary proficiency in the flying business, perceive the curious warping of the wings in the process of circling and the shifting position of the rudders. When after 1 minute 45 seconds of flight Wright again touched the ground, descending with extraordinary buoyancy and precision, while cheers arose from the crowd in the tribune, I saw the man who is said to be so unemotional turn pale. He had long suffered in silence; he was conscious that the world no longer doubted his achievements. . .

The Wrights had been terribly abused and defamed in France for no other reason than that there were people who did not want to believe that an American had beat them to the aeroplane. Thus, the account from France emphasized the need now for such unbelievers to surrender:

Mr. Wright has realised the most delicate problem of aviation— namely, the question of balance. To behold this flying machine turn sharp round at the end of the wood at a height of 60 feet, and continue on its course, is an enchanting spectacle. The wind does not seem to trouble him, Wright having flown in fairly stiff breezes. In a word, the Wright brothers are the first men who have succeeded in imitating birds. To deny it would be childish.

This is the sort of thing that appeared in newspapers (the preceding report was quoted in the *London Times*) a few months after Wittgenstein had arrived at the kite-flying station in Glossop, aiming to eventually design, build, and fly his own plane. In 1905, that the Wrights had flown was reported—and doubted even by some in aeronautical circles—but in the fall of 1908, their success, and their unexpectedly graceful and spectacular style of flying, were known to a much larger audience—to the general public, in fact. It was the most sensational news story of the season. Certainly the flight enthusiasts at the research station would have been enthralled by it. Thus, before Wittgenstein even began his first term as a research student in engineering at Manchester, there was universal recognition that the Wright Brothers' airplane was in a class by itself, way beyond what anyone else had achieved with their aircraft. The task of imitating bird flight had been accomplished.

Many French journalists granted even more: that the claims the Wrights had made in the past about their successful flights in 1903 must be believed. That meant the public demonstrations held recently and the prizes won by Farman and Santos-Dumont with their inferior performances were moot, as far as the question of priority went. The Wrights were recognized as "truly the first to fly" and Wilbur Wright as "the father of aviation."

The impact on England of the Wrights' visit and, especially, of Bleriot's crossing of the channel in an airplane the next year was almost immediate. In 1909, Britain began various initiatives to catch up with the Americans in aeronautics, including establishing the Advisory Committee on Aeronautics, on which Petavel, who ran the experimental station in Glossop, served.

Wilbur Wright stayed in Europe all that fall, astounding sold-out audience after sold-out audience. He took passengers up with him, handling the craft with the same sure control. All of European royalty decided they had to come and see it for themselves. All during Wittgenstein's first term at Manchester enrolled as an engineering research student, working with aeronautical enthusiasts, Wilbur Wright was in Europe exhibiting his airplane, and the press was lauding the Wrights as having solved the age-old problem of flight. Some (though not all) of the stubborn attitudes of envy and suspicion melted away when their holders saw Wilbur fly, and were instead replaced by pure exhilaration at seeing the Wright Flyer so gracefully and assuredly fly, quite literally, like a bird.

There was tragedy, too: In America, Orville's demonstration flights for the U.S. Army ended in a crash, killing Captain Thomas Selfridge and leaving Orville severely injured. Wilbur felt responsible, that he should have been there to make sure Orville was more careful. We saw in Gustav Lilienthal's memoir of his brother Otto the same sentiment. In Gustav's account, Otto crashed during one of the rare times he experimented without Gustav there. In Gustav's account, Otto was killed because he had not put the safety devices in place—the implication being that Gustav would not have let Otto fly without them, had he been there. Memory is unreliable, but it cannot be denied that these fatal flights occurred when brothers who usually worked together in experimental flights were apart.

It is no more surprising that Wittgenstein began looking for another momentous problem to solve than that he stayed on in Manchester after this and threw himself into aeronautical research projects. On the one hand, it must have been extraordinarily exciting to know that now humans could fly, to be living in a time when the problems of aviation one worked on could be expected to bear fruit in the form of better flying machines. It was a new age of flight, which many saw as a new age of mankind. How exciting to go to England to study aeronautical engineering and to be there for the birth of practical aviation, the creation of the British Advisory Committee on Aeronautics, the discussions about how to design reliable tests for the new and very advanced wind tunnel at the National Physical Laboratory! On the other hand, to a genius looking for the right use of his gifts, that the problem of practical flight had been decisively solved may have been an impetus instead to move on to another unsolved problem.

Wittgenstein biographers have noted that Wittgenstein next moved on to the problem of propeller design, devising and patenting a rather novel design. True, but seen in the context of the times, this refocusing of effort may have been more a response than a sign of initiative. In 1909, there was a request by the Admiralty for proposals for propeller designs for airships; Wittgenstein's patent for his propeller design is dated 1910. It seems to me he stayed at Manchester not so much to carry on any particular endeavor, but because there was no reason to leave. It was among the most comfortable places he had ever been. He could read and discuss whatever he liked, and that included the foundations of mathematics.

Yet, at some point, he did leave, and suddenly. The fall after his meeting with Frege in the first half of 1911, he went to Cambridge instead of returning to Manchester, where he was still enrolled as a research student. Whatever the reasons, the suddenness of the move was likely a more gradual outgrowth of his personality and interests, rather than a discontinuity necessitated by the situation in which he found himself. Many of the professors at Manchester had studied mathematics at Cambridge. Lamb, for instance, had placed second in the mathematical tripos, the competitive exam capping the course of studies in mathematics at Cambridge. For years afterward, his course of lectures preparing for it was one of the most sought-after there. Lamb's Hydrodynamics grew out of the lectures on the mathematical subject of ideal fluids that he developed there as part of a course of lectures to prepare for the tripos. Hence, Bertrand Russell's work on foundations of mathematics at Cambridge was not so alien to, nor distant from, the professors Wittgenstein encountered in the engineering department at Manchester as might at first appear. His interest has often been described as a progression from engineering to mathematics to foundations of mathematics, but it is doubtful that there was actually such a clear-cut linear progression in time of his interests. His multiple interests—in the problem of flight, in physics and mathematics, and in a theory of symbolism—had coexisted for a while. He had no reason to leave one of them behind for another: Boltzmann, whom he had so greatly admired, wrote about all three, as did Helmholtz. Many of the aeronautics researchers Wittgenstein encountered at Manchester, some of whom were physicists and mathematicians, had wide-ranging interests as well: In addition to work that would be regarded as pure scientific research, they employed the same knowledge in attacking practical engineering problems as the need arose. Yet, in terms of his personal destiny, it seems Wittgenstein did see a need to choose.

The move to Cambridge was a move in terms of where he thought his personal destiny lay. It is clear, however, that he thought he should "become a philosopher" only if he had the talent to do something great. In reading Russell's *Principles of Mathematics* and papers on foundations of logic, such as the papers presenting the theory of types, or in talking to Frege or reading his work, Wittgenstein was surely aware that the foundations of logic and mathematics were in a (slightly patched-up) crisis because of an unsolved problem. Frege thought the unsolved problem totally undermined his efforts to put mathematics on a firm logical basis, whereas Russell was optimistic about being able to circumvent the problem by placing some additional restrictions on the use of logical symbols.

The example the Wrights presented might have been influential. The Wrights were something of outsiders to the aeronautical scientific research community, yet, by first learning everything they could about what others knew, and then working in relative isolation from the aeronautical community, they had arrived in a matter of a few years at a strikingly elegant solution to a problem that others had made only slow and coarse stabs at. Boltzmann had cited Gauss's success in solving the problem of the constructibility of polygons as an analogy for the genius who would solve the problem of flight, but Gauss had not solved that problem by repeatedly taking a stab at constructing polygons. Instead, Gauss had shown how the problem in geometry was mirrored in algebra and solved it there by asking questions about the forms of algebraic equations.

Likewise, as was becoming known, the Wrights had solved the problem of flight, but not by an approach of repeatedly taking a stab at constructing better versions of a machine modeled on Pénaud's inherently stable design. The Wrights described a process of setting out to understand the effect of wind on differently-shaped wings. The breakthrough had come from constructing model wings they tested in their own wind tunnel; by these means, they then felt, they had found out the truth about the significance of differently-shaped wings for themselves. As for the problem that was stumping everyone, the problem of control, the Wrights totally redefined the goal. They started over, conceptually. The airplane would have a pilot to control it, as a bicycle had a rider to control it. The goal was not to design an inherently stable aircraft, but to design a method of using the instability of a plane to produce fine control with small, precise movements of the surfaces of the plane. In the process, they figured out what additional control surfaces would be required to make an airplane sufficiently responsive, and they reconceived the form of the airplane so that it contained such necessary control surfaces.

In retrospect, the Wrights' approach was as logical as their result was elegant. Perhaps hearing of their success in solving an age-old problem in such a logical way emboldened Wittgenstein to think that the problems plaguing the foundations of mathematics could be solved rather than merely coped with, and that it was something an outsider had as good a chance at doing as anyone else. He must have been building confidence in his abilities from the engineering work he was doing at Manchester as a research student, and this would have been happening while he was working on problems in logic on his own time. With the Wrights' solution to the problem of flight as the most salient example of the solution to an age-old problem that had been considered insoluble, one might be led to ask: Did the failures of others to solve Russell's paradox indicate that logic needed to be reconceptualized? Perhaps the symbolism used in logic was itself at fault. Russell had by this time presented a Mathematical Theory of Types to get around the paradox by placing certain restrictions on the use of symbols; types involved imposing a hierarchy of sorts on the logical symbolism. Boltzmann had talked about manipulating symbols and equations in the same breath as manipulating mechanical models. The question naturally arises: Did the mechanisms used in logic, the symbolism, need to be reformed, as the general conception of the airplane had?

Wittgenstein's sister describes a point in time at which he was "suddenly gripped by philosophy," "violently" and "against his will," and she dates it from his time at Charlottenberg "or a little later," which is consistent with late summer and early fall of 1908. Her account also indicates that he had already been thinking about philosophical problems at this point: "He was engaged in writing a philosophical work and finally made up his mind to show the plan of this work to a Professor Frege in Jena, who had discussed similar questions." Independent of this account, there is evidence that Wittgenstein thought he might have solved the paradox that had been plaguing Frege, and which had come to be known as "Russell's paradox"; McGuinness points out a relevant "scrap" of information in the mathematician P. E. B. Jourdain's notebooks, which records a conversation in 1909 with Wittgenstein and a parenthetical remark about him as follows: "Wittgenstein (who had 'solved' Russell's contradiction)."

Thus, although the move to Cambridge did not come about until he had spent three years at Manchester, it seems Wittgenstein was working on philosophy during that time. According to McGuinness: ". . . already in his first year in Manchester Wittgenstein had read enough to propose a contribution of his own, and it is probable that he had read some Russell before he met Frege. His later enthusiasm for both books leads one to suppose that he had read both Frege's *Grundgesetze der Arithmetik* and Russell's *Principles of Mathematics*." Besides the mathematician Littlewood, Manchester also had a professor of philosophy, Samuel Alexander, "who certainly knew the work of Frege and Russell and their definition of number." Alexander held Frege in very high esteem, and may even have recommended him to Wittgenstein explicitly.

But there came a time when Wittgenstein's response to the grip of philosophy became urgent: it was all or nothing. Either aeronautics or philosophy, not both; if he were to do philosophy, he must do it with all his might. Later, he described making other decisions in terms of a total commitment one way or the other—for example, choosing between the social life at Cambridge, which involved daily visits with colleagues, versus isolation from the whole university in a cabin in Norway. He often saw things like that when making decisions, even if he could not always keep the resolutions later.

The "all or nothing" approach ran through many aspects of his life, including his personal relationships. Though his friendship with William Eccles would be continuous from the time he first met him at the kite-flying station, many of the friendships he had after that were marked by sharp and emotionally wrenching discontinuities. The move to Cambridge was a sort of discontinuity, too: he arrived at the university there in the fall of 1911 without even applying or making arrangements of any sort, and while still enrolled as a research student at Manchester. Perhaps there was something tentative in it, too. He must have left behind some equipment and paraphernalia associated with his research projects in Manchester.

McGuinness writes that "His coming there had all the marks of an impulsive decision and of an experiment. Showing up at Cambridge might have been impulsive, but it was part of a longer-range plan on his part. We know that in 1911 he had written to Frege with some "objections to his theories," that Frege had warmly invited him to come to Jena, and that Frege told him he should go study with Bertrand Russell at Cambridge. Thus, we know that he didn't plunge into this "experiment" until several years after he had come across Russell's *Principles of Mathematics* and thought he had seen how to solve Russell's paradox.

The Wrights began their experiments in gliding about three years after news of Lilienthal's death had led them to start thinking anew about how they would go about solving the problem of practical flight. Wittgenstein, too, was experimenting here, but on himself rather than with an idea—how would *he* do at figuring out and solving the problems of logic? The place to try out his wings was wherever Bertrand Russell was, and in 1911 that was Cambridge, England. So he went there.

Chapter 5

A New Age-Old Problem to Solve

Wittgenstein's motivation in going to Cambridge was to attend Bertrand Russell's lectures, though he was to study psychology there as well. What Russell thought of him mattered a great deal to him at first. Russell reports that during his first term there, Wittgenstein pestered him to tell him what he thought of his abilities in philosophy: ". . . if I am a complete idiot, I shall become an aeronaut, but, if not, I shall become a philosopher." In Russell's record of the incident, he noted that Wittgenstein was "hesitating between philosophy and aviation," that he was "quite passionately interested in philosophy, but feels he ought not to give his life to it unless he is some good." Russell was not going to encourage someone just to provide encouragement, however. He asked Wittgenstein "to bring me something written to help me judge." It turned out well, and Wittgenstein did lay much upon the outcome; he later told David Pinsent that coming to Cambridge had "proved his salvation: for Russell had given him encouragement."

Cambridge was a wonderful place for Wittgenstein in terms of opportunities to hear music, too, and he made it part of his relationship with Russell. In giving a brief summary of that first year, McGuinness describes their friendship as "a great source of happiness to both" and writes: "Their discussions of philosophy were their chief work; each was stimulated by the other and confirmed by the other's reaction in the feeling that some advance was being made. They shared too nearly everything else that they thought important. Russell began to go to many more concerts, first taken by Wittgenstein" and tells of Wittgenstein's commenting to him "that hearing the Choral Symphony with Russell was one of the great moments of his life."

The other important friendship of Wittgenstein's first year at Cambridge developed through a shared appreciation of musical performances, too: his friendship with David Pinsent, to whom he later dedicated the *Tractatus*. "The two friends would borrow music from the Musical Union and take it to Pinsent's rooms. Eventually they developed a method of performing Schubert songs, Pinsent playing and Wittgenstein whistling (the truth and expressiveness of his whistling were often commented on). In this way, on holiday, they learnt 40 or 50 songs." Wittgenstein's first year there was a time of expanding horizons for Cambridge as well, for, as McGuinness's biography of him relates, the psychologist Charles Myers was in the process of building a new laboratory for experimental psychology at the time. It turns out the occasion was that Myers had come into an inheritance, and he used it to establish the Cambridge Psychological Laboratory, becoming its first director. It didn't open until 1913, but experimental psychology existed before then in "cottage rooms." For Wittgenstein, it would provide both a connection to topics he was familiar with from the intellectual milieu of his past days in Vienna and Berlin and another context for an important part of his new and future life at Cambridge: his friendship with Pinsent.

Although Myers later became interested in industrial psychology, he was at the time renowned for his research into musical perception, a legacy that remains in Cambridge today, judging from the university catalog:

> Science and music have been informally linked at Cambridge since the mid-nineteenth century. Sedley Taylor's 1873 volume "*Sound and music*," which included an account of "the chief acoustical discoveries of Professor Helmholtz," was intended to introduce his contemporaries to the new scientific insights on music emerging from Germany, while Charles Myers, later to found Cambridge's Experimental Psychology Laboratory, published pioneering research on nonwestern musical perceptions in 1905.

A "Science and Music" group at Cambridge is apparently thriving today.

Wittgenstein was studying experimental psychology as well as philosophy his first year in Cambridge, in 1911–1912. A great deal of the work on experimental psychology had been done by German-speaking psycho-physicists, including Mach and Helmholtz, two physicists who published popular works, and whom he almost certainly read in Vienna, and so were familiar to him. Experimental psychology laboratories were still extremely rare in England at the time. There may have been some continuity with Manchester here, too, for "Samuel Alexander, the philosophy professor at Manchester, had been committed to the new psychology. He had traveled to Freiburg to study experimental psychology with Munsterberg, during the academic session 1890–1891" writes Alan Costall, who also provides us with this surprising account from Winifred Hindshaw, one of the students in an advanced course in experimental psychology run by Alexander himself:

> I belonged to a small class doing Advanced Psychology with Professor Alexander (not to mention his famous terrier Griff) in the basement of the old building. We studied the special senses intensively, and did experiments relating to reaction times and attention. I think the apparatus in 1902-3 must have been rather simple. But we had a beautiful thing called a Plethysmograph which indicated pulse-changes, and therefore pleasure or pain according to the stimulus. . . . He was very interested in colour vision, and very precise about recording impressions. . . . He did not do hypnosis experiments with us, but he was interested in that side of inquiry. He had been to Nancy and knew about the work there.

D. H. Pear had been anointed to be Manchester's future psychologist even before he had finished his undergraduate degree, and it was Samuel Alexander's doing. He got Manchester not only to offer Pear support for undergraduate study, but to pay for him to learn German and travel to Germany to study with people there. This was all happening while Wittgenstein was at Manchester, and in 1911, the year Wittgenstein left Manchester for Cambridge, Pear published two papers on the psychology of music.

Wittgenstein's psychology experiments likewise had to do with musical perception, and he asked David Pinsent to be a subject. Myers gave a talk on "Primitive Music" in February of 1912 at Cambridge, which

Wittgenstein must have heard of, and probably attended. According to McGuinness, Myers' view was that various hypotheses about the development of music notwithstanding, "it was better to think that both language and music developed out of a more primitive system of communication. . . . The main application that Myers could make of the study of primitive music was the isolation of the various factors in music appreciation as we know it . . ." He brings special attention to one of these factors that may have been important to Wittgenstein: "Musikgestaltqualitat—the property of being musically meaningful, of constituting a phrase or tune. This last distinguishes music from mere noise and Myers describes interesting cases of loss of the sense for it—sometimes by highly musical people, who could still distinguish all the differences in what they heard but were unable to hear it as music."

In Chapter 1, I speculated that the gramophone, since it appeared just in time for Wittgenstein's birth, and since he lived in a household in which musical performances were constantly discussed, would naturally raise questions about representation of musical performances, and that such questions would be related to questions of representation of the performance of models of aircraft by the toy helicopters so popular in Europe at the time. The gramophone does appear in a key passage in the *Tractatus*—the passage on what logical form is—and in many of his writings afterwards. The work in experimental psychology on music perception reinforced, and may have stimulated, further questions about how language and music are alike and how they are not alike. By the time he wrote the *Tractatus*, Wittgenstein had integrated the ideas about representation by gramophone recordings with the ideas about meaning in music and meaning in language; they appear interwoven when, in explaining what he means, he says "A proposition is a picture of reality. A proposition is a model of reality as we imagine it." The analogy is used in thinking about symbolism:

> 4.011 At first sight a proposition—one set out on the printed page, for example—does not seem to be a picture of the reality with which it is concerned. But neither do written notes seem at first sight to be a picture of a piece of music, nor our phonetic notation (the alphabet) to be a picture of our speech. And yet these sign-languages prove to be pictures, even in the ordinary sense, of what they represent.

In Wittgenstein's first year at Cambridge, he and Pinsent spent many hours in experiments and in discussion about the topic being researched: rhythms. The topic reflects Myers' interests; it may or may not have reflected Wittgenstein's. At any rate, Wittgenstein did produce a professional paper on the subject and presented it at the British Psychological Society in the summer of 1912, though he described it to Russell as "absurd." At the opening of the Cambridge Psychological Laboratory in 1913, he demonstrated an apparatus used for associated psychological experiments. He was later to reflect on the relation of psychology and philosophy in the *Tractatus*. Psychology was not for him, as it was for some, the area of philosophy called theory of knowledge; rather, "Theory of knowledge is the philosophy of psychology." He felt that there was the constant danger of losing one's way in doing experimental psychology (as with doing any other natural science): "Does not my study of sign-language correspond to the study of thought-processes, which philosophers used to consider so essential to the philosophy of logic? Only in most cases they got entangled in unessential psychological investigations, and with my method too there is an analogous risk." Here Frege's view would certainly have mattered to him—provided a sort of beacon—for Frege was emphatic about keeping clear about the distinction between the psychological and the logical. During this time, Wittgenstein worked with the mathematician Philip Jourdain on a translation of Frege's *Grundgesetze*, or *Basic Laws of Arithmetic*.

The results of the experiment with himself in coming to Cambridge and presenting himself to Russell were positive: he could, and would, become a philosopher. He had a problem to think about: the problem of finding a symbolism that would not require the kind of ad hoc restrictions Russell had come up with in order to escape the paradoxes. Russell and Whitehead's logical symbolism, as with most everyone else's, included "logical constants"—the symbols in logic that stand for connectives such as "if . . . then," "or," and "and." Wittgenstein had a strong conviction that the right symbolism would not contain any such logical constants. In a letter to Russell, in which he mentions all in a row without a breath between, that he has had a discussion with Myers about the relationship between logic and psychology in which he was "quite wilde," that he reads William

James' "Varieties of Religious Experience" whenever he can, and the sort of good that does him, he writes:

> Logic is still in the melting pot but one thing gets more & more obvious to me: The props of Logic contain *only apparent* variables & whatever may turn out to be the proper explanation of apparent variables, it's co[n]sequence must be that there are *no* logical constants.

Here Wittgenstein is referring to how Russell distinguishes between "apparent" variables, which is the use of a variable within the scope of a quantifier such as "any" or "all" (now often referred to as "bound" variables), as opposed to free variables. Wittgenstein's conviction here echoes the view Frege had clarified in discussions about the axioms of geometry in his (public) correspondence with Hilbert. Frege there pointed out that the sentence "All x such that x is a square root of 4 is even," or, alternatively, "If x is a square root of 4, then x is even" (in which, in Russell's terms, x is only an apparent variable) is a genuine proposition, for the statement is complete without needing anything to be substituted for x. However, the statements "x is a square root of 4" and "x is even" are *not* propositions, for they require supplementation to be made into propositions; Frege called these "pseudo-propositions." In saying that "The props of Logic contain only apparent variables," Wittgenstein is endorsing Frege's view about genuine propositions versus pseudo-propositions. The statement that "whatever may turn out to be the proper explanation of apparent variables, it's co[n]sequence must be that there are no logical constants," however, is Wittgenstein's contribution. In fact, he would later say that the *Grundgedanke*, or fundamental thought, of the *Tractatus* was that there were no logical constants.

Russell was dissatisfied with his theory of types, but he hoped to make it work. Wittgenstein, on the other hand, thought any theory of types was doomed to fail. In early 1913, he wrote to Russell from Vienna, where he was staying due to his father's terminal illness, of some new ideas he had, and that:

> What I am most certain of is . . . the fact that all theory of types must be done away with by a theory of symbolism showing that what seem to be different kinds of things are symbolized by different kinds of symbols which cannot possibly be substituted in one another's places.

In a postscript he adds:

> Propositions which I formerly wrote (a, R, b) I now write R (a,b) & analyse them into a, b, & ($x, y) R (x,y).

What Wittgenstein is saying in this postscript is that, instead of an ordered triplet in which only things of the right type can be substituted for each of the symbols in the triplet (a and b by individuals and R by relations between those types of individuals), he uses R (a,b) instead, and dispenses with restrictions on what can be substituted for a and b. He was not certain of the specifics of his analysis, but his conviction that the theory of types must be abandoned was firm. The difference between the two formalisms might not seem so great, but it reflected a fundamentally different view of symbols.

It was as fundamental a difference in logic as the difference between Pénaud's use of inherent stability and the approach the Wright Brothers had taken in rejecting the presumption that an airplane ought to be inherently stable. Inherent stability comes at a cost: maneuverability—the pilot's ability to use any means of control, about any axis, that was physically possible.

The reason for Wittgenstein's conviction that Russell's basic approach to a correct theory of symbolism is totally misguided is clearer in a letter he wrote later, after the *Tractatus* was written. Responding to Russell's statement in 1919 that:

> The theory of types, in my view, is a theory of correct symbolism:
> (a) a simple symbol must not be used to express anything complex;
> (b) more generally, a symbol must have the same structure as its meaning.

Wittgenstein writes back:

> That's exactly what one can't say. You cannot prescribe to a symbol what it may be used to express. All that a symbol can express, it may express. This is a short answer but it is true!

The conviction that, on a correct theory of symbolism, there would not be any logical constants such as symbols for "and" and "or" in the most general form of a proposition is not unrelated, for logical constants are stipulations about what symbols may represent as well. Just as Wittgenstein admonishes Russell against putting restrictions on what a symbol may express, so the Wright Brothers preached against the general approach of building an inherently stable aircraft, since building inherent stability into the airplane's design was a restriction on methods of controlling the aircraft. An airplane can move about three axes: rotation about the axis that runs the length of the plane from front to back is roll (one wing goes up the other down), rotation about the axis that runs the length of the wings is pitch (the nose of the plane goes up and down), and rotation about the plane's vertical axis is yaw (turning from side to side, from left to right). The vice was in adding restrictions. No matter if these restrictions could help avoid certain catastrophic events, they were not the way to the correct method of aircraft control.

Russell had developed the theory of types for analogous reasons: to prevent the kind of catastrophic uses of symbolism that were exemplified by the paradoxes that bear his name, though in truth they dated from the time of the Ancient Greeks. He explains the motivation for types in the popular treatment "The Logical Theory of Types" about 1910, which shows his view just before he met Wittgenstein:

> It is agreed that the paradoxes to be avoided all result from a certain kind of vicious circle. The vicious circles in question all arise from supposing that a collection of objects may contain members which can only be defined by means of the collection as a whole. Thus, for example, the collection of propositions will be supposed to contain a proposition stating that "all propositions are either true or false." It would seem, however, that such a statement could not be legitimate unless "all propositions" referred to some already definite collection, which it cannot do if new propositions are created by statements about "all propositions." We shall, therefore, have to say that statements about "all propositions" are meaningless. . . . Propositions, as the above illustration shows, must be a set having no total. The same is true . . . of propositional functions, . . . it is necessary to break up our set into smaller sets, each of which is capable of a total. This is what the theory of types aims at effecting.

The connection between the theory of types is explained there, too:

> The paradoxes that more nearly concern the mathematician are all concerned *with propositional functions.* By a "propositional function" I mean something which contains a variable x, and expresses a proposition as soon as a value is assigned to x. That is to say, it differs from a proposition solely by the fact that it is ambiguous: it contains a variable of which the value is unassigned. . . . Thus e.g. "x is a man" or "sin x = 1" is a propositional function. We shall find that it is possible to incur a vicious-circle fallacy right at the very outset by admitting as possible arguments to a propositional function terms which presuppose the function. This form of the fallacy is very instructive, and its avoidance leads, as we shall see, to the hierarchy of types.

Thus, Russell avoids the logician's most-feared catastrophe, fallacy, by categorizing the terms that can be substituted for certain symbols and by laying out restrictions for how symbols can be used by specifying types. At the cost of restricting the ability to make use of all the possible uses a symbol may have, Russell provided a kind of inherent stability (an avoidance of fallacy and paradox) in his logical system. It was not built into the symbolism itself, but into restrictions on how the symbols could be used. It is somewhat like a theory of the airplane that aimed for inherent stability and restricted the kinds of movements available to the pilot.

The notion of inherent stability is not the same as the notion of automatic stability. In automatic stability, the airplane has stabilizing auto-mechanisms so that it can "take care of itself" (as Wittgenstein thought logic must take care of itself) while going in arbitrary directions as directed by the pilot. The Wrights were not against automatic mechanisms to ensure stability; the issue wasn't a matter of human involvement in favor of the machine's taking care of itself, of conscious intent versus unconscious or mechanical habit. Their objection was to the notion of inherent stability, the approach of restricting the freedom of the aircraft to move in certain ways in order to prevent it from getting into an unstable situation where its path could be catastrophically undone by the wind or other forces. (Think of the difference between a tricycle and a bicycle.) Rather than designing an inherently stable aircraft and wondering how it could be clumsily nudged to change direction within the limited opportunities for control provided by the plane's design, the Wrights instead allowed their airplane to be unstable and asked what kind of additional control surfaces were needed, so that, like riding a bicycle, a well-trained pilot could make the plane go just where he wanted it to go, when he wished.

What distinguished the Wright Brothers' approach from all the mentors and colleagues with whom they discussed flying machines was their rejection of inherent stability, and they rejected inherent stability because it put restrictions on some of the means of maneuverability. What distinguished Wittgenstein from his mentor and colleague Russell, with whom he discussed theories of symbolism, was his rejection of the theory of types, and he rejected the theory of types because it put restrictions on some of the means a symbol in a proposition may be used to express. As the Wrights' approach relied on a pilot's skill, or a pilot's skill in conjunction with automatic mechanisms, so Wittgenstein's account of sign and symbolism relies on human skills—such as the ability to read a musical score.

Wittgenstein's work at Cambridge was interrupted by some visits home occasioned by his father's illness; his father died in early 1913. He spent time with various friends, family, and acquaintances that year, but in late 1913, he began making plans to withdraw to Norway, away from Cambridge, away from people, to work on solving the problems of logic. Russell was mentally tired and worn out from years of intense work on *Principia Mathematica*, and he was very glad to have the bright, young, intense Wittgenstein to hand the baton to. They worked on problems together by now, but more and more it was Wittgenstein who was judging and Russell who looked to Wittgenstein's judgement.

There was no looking back at aeronautics as an alternative career anymore. Wittgenstein had made the agonizing decision to become a philosopher, and he had found the age-old problem that he felt he was meant to solve: finding a correct theory of symbolism.

Chapter 6

The Physics of Miniature Worlds

Wittgenstein's investigations into logic were bringing him around to notions of mirroring and corresponding. In notes expressing his views as of April 1914, he concludes "Thus a language which *can* express everything *mirrors* certain properties of the world by these properties which it must have." And he struggles to accommodate his observation of the problematic fact that "in the case of different propositions, the way in which they correspond to the facts to which they correspond is quite different." A year earlier, he had said there was no such thing as the form of a proposition ("the form of a proposition is not a thing"); his resolution of the issue of how propositions correspond to facts now is in terms of the *general* form of a proposition—something, he has decided, that all propositions *do* have in common. Then he says, "In giving the general form of a proposition, you are explaining what kind of ways of putting together the symbols of things and relations will correspond to (be analogous to) the things having those relations in reality."

These exploratory thoughts about the notion of correspondence were the beginning steps toward an answer to one of the puzzles raised much earlier by musical scores and the gramophone records that had been such a striking arrival on the scene the year Wittgenstein was born: "What is the relationship between the symbols in the score and the patterns of grooves in the gramophone record?" That there was a mechanical process that could be used to make a gramophone record and one that could be used to play sound from it was well known. What about the process of creating a musical score, and the process by which a symphony could be imagined or produced by a musician reading the score? Were these just as straightforward? Wittgenstein had already steered clear of simplistic accounts of a symbol as "sign of thing signified" a year earlier, at least in the case of words, in deciding that "Man possesses an innate capacity for constructing symbols with which some sense can be expressed, without having the slightest idea what each word signifies." But how did mirroring work, if not by a straightforward correspondence?

Wittgenstein was not alone in pondering how items of language could mirror a situation and how propositions could correspond to the world. The specific suggestion that equations function like pictures or models was made by Boltzmann in his *Lectures on the Principles of Mechanics*, a work in which he strove for an accurate exposition of mechanics that would be accessible to members of the general public. In explaining the role of pictures in physical theories, Boltzmann had there explained that even those who thought their approach had dispensed with pictures had not really done so: "[Partial differential equations] too are nothing more than rules for constructing alien mental pictures, namely of series of numbers. Partial differential equations require the construction of collections of numbers representing a manifold of dimensions." Thus, he said, at the bottom Maxwell's equations "like all partial differential equations of mathematical physics . . . are likewise only inexact schematic pictures for definite areas of fact." Boltzmann's suggestion, however, went just as far as claiming that symbolic equations could function like scientific models or pictures—it did not purport to explain exactly how either worked. It does seem, though, that picturing involved some imagined entities that may, but need not, correspond to something in reality. He speaks of pictures almost interchangeably with mental pictures.

Boltzmann became an extremely popular lecturer in Vienna around 1903, when, as mentioned earlier, Wittgenstein would have been about fourteen years old and would have known of Boltzmann's lectures. These lectures were so popular that the lecture hall in Vienna could not accommodate the audience, and Boltzmann was invited to give them at the palace instead. Boltzmann was present in Wittgenstein's youth through his prolific writings as well as through these lectures delivered a stone's throw from his home. The second volume of Boltzmann's *Lectures on the*

Principles of Mechanics was published in 1904, and a collection of his writings was published as *Popular Writings* in 1905, when Wittgenstein was sixteen. As we have seen, at that crucial time in his life, he was so interested in Boltzmann that, at least as he later recounted things to his friend and colleague von Wright, he had originally planned to study physics with him.

Boltzmann's *Popular Writings* anthology included an essay republished from a physics journal, "On the Indispensibility of Atomism in Natural Science," in which he emphasizes the remarks about equations quoted earlier: "The differential equations of mathematico-physical phenomenology are evidently nothing but rules for forming and combining numbers and geometrical concepts, and these in turn are nothing but mental pictures from which appearances can be predicted." Here he refers the reader to Ernst Mach's *Principles of the Theory of Heat*, remarking that Mach's work has helped him clarify his views. Boltzmann here stresses that, as far as the use of models goes, there is no essential difference in the approach he takes and approaches such as energetics, in which equations rather than models of material points, are central: "Exactly the same holds for the conceptions of atomism, so that in this respect I cannot discern the least difference. In any case it seems to me that of a comprehensive area of fact we can never have a direct description but always only a mental picture." He had a rather precise criticism specific to differential equations and the corresponding assumption of a continuum, in that differential equations relied on the notion of a limit, and that observationally there was no distinguishing between systems of large numbers of finite particles and actual continuums. Thus, he said "those who imagine they have got rid of atomism by means of differential equations fail to see the wood for the trees."

Elsewhere, in his encyclopedia article on "Model," which was reprinted in the same anthology, Boltzmann again described the method he referred to as the theory of "mechanical analogies," remarking that, unlike in earlier days, "nowadays philosophers postulate no more than a partial resemblance between the phenomena visible in such mechanisms and those which appear in nature." Looking closely at his remarks, though, it is clear he had run into a brick wall with this approach. He had to except from the models to which his remarks applied the kind of model that was used in experimental engineering scale models. On the approach in which physical models constructed with our own hands are actually a continuation and integration of our process of thought, he says, "physical theory is merely a mental construction of mechanical models, the working of which we make plain to ourselves by the analogy of mechanisms we hold in our hands." In contrast, in his discussion of mental models, Boltzmann had explicitly described experimental models as of a different sort than the kind with which he was comparing mental models. Boltzmann even explained why they must be distinguished:

> A distinction must be observed between the models which have been described and those experimental models which present on a small scale a machine that is subsequently to be completed on a larger, so as to afford a trial of its capabilities. Here it must be noted that a mere alteration in dimensions is often sufficient to cause a material alteration in the action, since the various capabilities depend in various ways on the linear dimensions. Thus the weight varies as the cube of the linear dimensions, the surface of any single part and the phenomena that depend on such surfaces are proportionate to the square, while other effects—such as friction, expansion and condition of heat, etc., vary according to other laws. Hence a flying-machine, which when made on a small scale is able to support its own weight, loses its power when its dimensions are increased. The theory, initiated by Sir Isaac Newton, of the dependence of various effects on the linear dimensions, is treated in the article UNITS, DIMENSIONS OF.

Thus, the experimental models represent a challenge: for experimental models, the relationship between model and what is modeled is in some ways unlike the relationship between a mental model and what is modeled by it.

Boltzmann committed suicide in 1906, the year after the anthology appeared, and Wittgenstein never did get to study with him. This remark of Boltzmann's might well have resonated with Wittgenstein's personal experience, even though he did not get to do experimental work under Boltzmann, for we know that Wittgenstein had built and played with a toy airplane, and these toys were quite serious affairs technically. Boltzmann's remarks about experimental models, and his specific mention of a model of a flying machine as a model that does not behave like the full-size machine it models, could scarcely fail to command Wittgenstein's attention. If an airplane design would only work the same way when enlarged, the problem of sustained, controlled heavier-than-air flight would have pretty much already been in the hands of countless children in Europe and America. As we saw earlier, Pénaud had developed a rubber band-powered model airplane that was capable of sustained, stable flight, and some of Pénaud's designs were available as toys even before Wittgenstein was born. Boltzmann was especially aware of the fact that, in England, Maxim had shown that it was possible to design a full-size steam-powered airplane capable of getting off the ground. In doing so, Maxim showed that an airfoil or kite could be powered—that is, that the power an engine produced could be large enough in proportion to its weight to get an airplane off the ground, which is all he was trying to establish at that point. The unsurmounted obstacle was to get the sustained flight that Pénaud had already achieved in a small-scale model, in a full-size airplane. Pénaud's work had been the most promising, but as we saw earlier, he too had committed suicide. In fact, he had done so upon receiving news that construction of the full-size model he had designed would not receive the funding he had been expecting. In 1905, the air was full of the promise of controlled heavier-than-air flight, and there were some who believed the stories that two Americans had achieved it. Thus, Boltzmann's remarks describing significant and essential differences between mental models and models of flying machines would have had the effect of diminishing interest in mental models, because they made mental models seem like less robust representations of the world. And they may well have piqued interest in understanding whether and how models of flying machines could represent larger ones.

The puzzlement about the effect of size on the ability of machines to fly was common to just about anyone who played with toy flying machines. We saw earlier that the Wright Brothers recalled very clearly their puzzlement as children that the larger-sized models they built of exactly the same design didn't perform like the wonderful toy did. Even had Wittgenstein at age sixteen not recalled similar experiences with his childhood toys when reading this passage in Boltzmann's 1905 anthology, Boltzmann's point about the effect of size on the strength and performance of machines would almost certainly be remembered, given that Wittgenstein soon found himself enrolled in an engineering certificate program, and especially interested in aeronautics.

Certainly by the time he left his position as a research student in aeronautical engineering at Manchester in the fall of 1911 to show up at Cambridge asking to study logic with Bertrand Russell, Wittgenstein would have been familiar with the use of experimental scale models for specific types of engineering problems. Two wind tunnels were already in use for aeronautical research in England by that time. There was the tunnel that Wenham had convinced the Aeronautical Society of Great Britain to fund before anyone fully understood in general how to use the results on small-size models to predict the behavior of a full-size model. Then there was the privately-funded wind tunnel that the wealthy inventor Maxim had built, which was constructed after Reynolds had shown how to use experiments to predict behavior in similar flow situations—at least for fluid flow in pipes. The significant thing about Reynolds' work was that it provided a way to determine similarity of flow regime for different-sized pipes, as well as for flow at different velocities and viscosities. Reynolds' work used liquids to investigate fluid flow, but air is a fluid, too, so Reynolds' work also bore on the questions of how wind tunnel results on models of aerodynamic surfaces could be used to predict the behavior of larger versions of the surfaces tested, and how that behavior would vary with different air velocities.

This is not to say that Wittgenstein would have then known exactly how to pick up where Boltzmann left off and fill in the story for experimental models, for it is not clear that there was an account of a general methodology of experimental models at that time. The way things stood with the practice of using engineering scale models might well have evoked almost exactly the puzzlement Wittgenstein had about propositions: "in the case of different propositions, the way in which they correspond to the facts to which they correspond is quite different." Given the approach then to experimental engineering models, where the rules about how to scale from results on a model to results on a full-size object depended on whether you were talking about experiments in a towing canal or fluid flow in a pipe, one could just as well say that, in the case of different models, "the way in which they correspond to the things they model is quite different."

What would the analogous point he had made for propositions be for experimental models? Wittgenstein's view in early 1914 about propositions was that "In giving the general form of a proposition you are explaining what kind of ways of putting together the symbols of things and relations will correspond to (be analogous to) the things having those relations in reality." What could it mean to give the general form of a model? Or, on Boltzmann's view that equations are really models of a sort, what does it mean to give the most general form of an equation?

In early 1914, Wittgenstein was asking these questions for propositions. Curiously, as we shall see, by the end of 1914, there would be a paper in the field of physics addressing analogous questions about empirical equations. The investigation in that physics paper involved finding the general form of an empirical equation, and it ended up addressing the question of what a universe built on a smaller scale would be like. There was more to be said in answering the question about the relationship between empirical equations and models than Wittgenstein was able to say about propositions in early 1914, the extra twist having to do with the fact that empirical equations involve measurement. The answer given for such equations would appear in late 1914 in a paper that also presented a formal basis for the methodology of experimental models. Though its author, Edgar Buckingham, was American, he had studied in Leipzig with Wilhelm Ostwald for his doctorate and had written a book on the foundations of thermodynamics. Thus, Buckingham's discussion was informed by the debates between Boltzmann and Ostwald about energetics, the kinetic theory of gases, and statistical thermodynamics.

Boltzmann had tried to tone down the strident claims of supporters of energetics such as Ostwald, who was antagonistic to the use of models. Ostwald's view, at least as Boltzmann understood Ostwald's emphasis, was not only that the use of models in thermodynamics and the kinetic theory of gases were so much extraneous and distracting baggage, but also that the use of models at all was suspect. In defending the use of models against such strident claims, we saw, Boltzmann pointed out that even proponents of energetics used models of a sort, inasmuch as they used equations as a sort of model—a model made of symbols. Boltzmann's suggestion that equations function like models may well have prodded Wittgenstein to think of a proposition as a model, and it may have even been implicit in some of Wittgenstein's statements in the manuscripts on logic he was working on in 1913 and 1914. At any rate, the notion of a proposition being like a model in some way was not explicit then. Wittgenstein just did not talk about propositions being models or pictures during his stay in Norway—that would come only after the crucial insight in late 1914.

However, in early 1914, Wittgenstein *was* talking about propositions in terms of the facts to which they correspond, as was Russell. In the first manuscript on logic he produced in 1914, he writes "Propositions [which are symbols having reference to facts] are themselves facts: that this inkpot is on this table may express that I sit in this chair." Wittgenstein's move here about propositions and facts is at least vaguely reflective of Boltzmann's move in saying that manipulating symbols in an equation is using the equation like a model. Likewise, as indicated in the preceding quote from his manuscript, Wittgenstein had already, during his time in Norway, made the move that propositions not only are symbols that correspond to facts, but are themselves facts—that is, they are the same sort of thing that they correspond to. Thinking of propositions as facts was not new. Frege had spoken of the marks on paper associated with a written sentence, and Russell of the varieties of facts that correspond to a proposition, including the example of the acoustic fact associated with a spoken sentence. Frege and Russell, though, tended to de-emphasize this kind of fact, to mention it only by way of contrasting the kind of fact that a proposition *is* with the kind of fact to which a proposition *corresponded*. But, in all fairness, even if Wittgenstein was tentatively exploring the possibilities of the observation that a proposition is a fact, rather than merely mentioning it as a contrast to the kind of fact to which a proposition does correspond before the war, he was not exploring the idea of a proposition being a fact *in terms of picturing or modeling* then. The key notion regarding propositions that shows up in the work Wittgenstein did during his stay in Norway just before the outbreak of World War I is the notion of *correspondence* rather than the notion of *picture* or *model*.

It was almost as though something in the atmosphere was stimulating people's appetites for a satisfying understanding of correspondence, similarity, and form. For, while Wittgenstein was living in Norway pondering problematic issues in logic such as the fact that "it seems as if, in the case of different propositions, the way in which they correspond to the facts to which they correspond is quite different," interest in similarity, correspondence, and similarity transformations was appearing in a wide variety of contexts, especially in Britain.

Someone familiar with Boltzmann's *Popular Scientific Writings* might well find that these discussions of different kinds of correspondence, and especially the accompanying explorations of the consequences of similarity, brought to mind Boltzmann's remark about the methods of theoretical physics: "The new approach compensates the abandonment of complete congruence with nature by the correspondingly more striking appearance of the points of similarity. No doubt the future belongs to this new method." Boltzmann had written that in 1892, in the wake of Hertz's spectacular successes in electrodynamics, which in turn (according to Boltzmann) owed much to Maxwell's ingenious mechanical analogies for his equations describing electromagnetic phenomena.

Boltzmann's point here is that, although the analogies Maxwell came up with were crucial to Hertz's advances, Maxwell did not intend the analogies to be taken literally. Maxwell did not mean them to be regarded as hypotheses; Boltzmann felt that, just as with Maxwell's attitude toward his own equations, so things had become in general with the equations of theoretical physics: as he put it there, science speaks "merely in similes." Two decades after Boltzmann penned these remarks about the new methods in physics, we find that in the contexts of discussing thermodynamics, hydrodynamics, and biology (morphology), many other thinkers in many other fields were seeking definitive statements about similarity, too. Not all the ideas that sprang up were unprecedented, but ideas about the use of similarity, whether old or new, were now being explicitly reflected upon, talked about, and written about. In the years 1913–1914 in particular, simultaneous activity of this sort occurred in a number of very different disciplines. Looking back, the activity in the few years just prior to 1914 portends a convergence of ideas about similarity and correspondence.

Certainly some notion of correspondence was already familiar, from pure mathematics as well as from theoretical mechanics. Bertrand Russell's *Principia Mathematica*, published in England in three volumes from 1910 to 1913, reflected the approaches of pure mathematicians in Germany, such as Richard Dedekind and Georg Cantor. Dedekind and Cantor, friends and colleagues, became friends and published very different definitions of the real numbers in 1872, but both had employed the notions of correspondence and similarity in formalizing and defining numbers and other mathematical concepts. Dedekind, for instance, had set up an analogy between the set of rational numbers and a straight line, which he then used to explicitly define a correspondence between rational numbers and points on the line. Rational numbers are numbers that can be expressed as the ratio of two whole numbers; irrational numbers cannot. Constructing a line segment corresponding to a rational number on a line uses only the most basic methods of geometrical construction, so Dedekind considered the rational numbers a logical starting point from which to define the rest of the real numbers. This correspondence was not merely illustrative, but was put to good use, for the fact that there were some points on the straight line to which no rational numbers corresponded motivated a definition of irrational numbers. Numbers, as Dedekind defined them, corresponded to "cuts" of the rational numbers that were analogous to "cuts" of the straight line. So analogy and correspondence were not of merely heuristic value in discovering the definition of number, but vestiges of them actually appear in the definition. Dedekind also showed how to use the definition of real numbers to construct a definition of complex numbers (numbers of the form a + bi, where a and b are real numbers and i is the square root of −1).

Cantor took a totally different approach. He did not presume the rational numbers, or any kind of numbers, as already familiar and known, and he aimed for a notion of number even more general than real numbers and complex numbers. Cantor's approach was to start with what he thought one of the most basic of mental activities—abstracting from individual properties of objects in a collection. This resulted in a definition of what he called "transfinite" numbers, which contrast with "finite" numbers. He gave a definition of infinite that distinguished infinite sets and finite sets, and in doing so discovered a whole world of infinite numbers. His definitions are based on two main notions: the notion of one-to-one correspondence and the notion of an order-preserving mapping. Roughly put, the notion of one-to-one correspondence captures the idea of how many, whereas the notion of an order-preserving mapping captures the notion of similarity of order structure. He found that these notions led to a way to delineate infinite sets. His definition of infinite says that infinite sets are sets that can be put in one-to-one correspondence with a proper subset of themselves, something that is impossible for finite sets. On this definition of infinite, the set of whole numbers is easily seen to be infinite because it can be put into one-to-one correspondence with the even numbers (just map each number n to the number $2n$), and the even numbers are contained within the set of whole numbers. He defined cardinal type, which expresses the informal idea of how many things are in a collection, and ordinal type, which expresses the informal idea of how things are ordered in a collection as well as how many things are in it. As he defined cardinal number and ordinal number, the two concepts coincide for finite numbers but are forced apart for transfinite (infinite) numbers. Russell's third volume of *Principia Mathematica*, published in 1913, dealt with transfinite cardinal and ordinal numbers and incorporated many of Cantor's ideas. Thus, these ideas became more widely known in England in 1913, especially among philosophers and people interested in philosophy and logic.

The ideas were the horizon-expanding kind, provoking the kind of exhilaration in thought that ballooning had for the senses. They made people feel that, using only simple, familiar ideas, they were transported to a realm where they were suddenly freed of things that had bound them before. Similarity was shown to be a very powerful, if unassuming, idea. Cantor argued that not only had he extended the notion of number, but that the notion of a transfinite ordinal number reflected the most general notion of number possible. The notion of "similar" aggregates (he used the term "aggregates" to refer to collections formed by the mind by abstracting from individual characteristics of the things in the collection) turned out to be exceedingly fruitful. It required a notion of something more structured conceptually than a mere aggregate. Thus, Cantor was led to define ordered aggregates (where different notions of "less than" induce different orders) and then an even more important concept: a well-ordered aggregate, which has a "least element"—a starting point or end point. We can think of one aggregate being "transformed" into an aggregate to which it is similar by a similarity transformation, just as we can think of an aggregate being transformed into one to which it is equipollent (meaning that its elements and the transformed one can be put into a one-to-one correspondence) by a simple replacement of each element of the aggregate, one by one. The mathematical notion of a transformation can be used to discuss a mapping. Instead of describing a mapping, which involves specifying a rule by which each element in one aggregate or object is paired with an element in the aggregate or object to which it is similar, we can talk about an aggregate or object being transformed into one to which it is similar.

Similarity is not just a mathematical notion, however. The notion of similarity is entwined in the thought and practice of just about any discipline you can think of, although it is not always talked about per se. Then, as now, notions of similarity were essential to much of scientific reasoning and engineering practice, even though discussion of the topic itself appears infrequently in scientific papers. However, in the years just prior to and including 1914, there was a cultural precipitation of papers that did explicitly reflect on the use of similarity.

In 1912, when James Thompson's *Collected Papers in Science and Engineering* appeared (twenty years after his death), it contained a paper about similar structures. These are structures in the most literal sense—structures such as bridges and columns. In that paper, "Comparison of similar structures as to elasticity, strength, and stability," he distinguishes two kinds of similarity between structures: similarity with respect to elasticity and bending, and similarity with respect to stability. The paper was written in 1875, just after Dedekind and Cantor published their accounts of number. James Thompson's style of reasoning illustrates that, in practical engineering, the method of similarity, though based on reasoning from principles of natural science, was still conceived of in terms of *specific kinds* of similarity, specific kinds of loads (wind on a surface versus attached weight), and specific disciplines (hydrodynamic versus mechanics of materials). The kinds of things that were then called "similarity principles" were statements covering a certain class of cases. The point of the "principle" was usually to state how one variable—the weight, size, elasticity—was to be varied as the linear dimension was varied—that is, as an object was increased or decreased in size but kept the same shape. James Thompson's examples are often about how to vary some quantity such that two structures of different sizes are similar in one of these respects. Here is one example of what is meant by a "similarity principle" taken from that work: "Similar structures, if strained similarly within limits of elasticity from their forms when free from applied forces, must have their systems of applied forces, similar in arrangement and of amounts, at homologous places, proportional to the squares of their homologous linear dimensions." Sometimes the reasoning is based on equations, but often it is not. Rather, some arguments from physical intuition are used: that weight increases as the cube of the linear dimension, and a cross-sectional area of a rope increases as the square of the linear dimension. They are very much like the statements, cited earlier, that Boltzmann used in describing the kind of model he called "those experimental models which present on a small scale a machine that is subsequently to be completed on a larger, so as to afford a trial of its capabilities."

At about the same time, the polymath D'Arcy Wentworth Thompson, then a professor of biology at the University of St. Andrews, was working out ideas that would soon appear in a compilation titled *On Growth and Form*. It is most well-known for its illustrations showing sketches of animals and animal parts that are "morphed" into others. Each sketch is overlaid with a grid; in the transformed sketch, the grid is stretched or slanted in some way so that the form of one species of animal looks as though it is obtained from another via a transformation mapping the points on the lines of one sketch to another by a mathematical function. Some commentators today regard this work as putting forth an alternative to Darwin's theory of natural selection, but this obscures the nature of D'Arcy Thompson's masterwork. Certainly it is true that D'Arcy Thompson wanted to put the brakes on the tendency of his contemporaries to use Darwin's theory of natural selection to explain everything, to the exclusion of other kinds of explanations. But careful readers of Darwin know that scientific explanations of animal forms according to Darwin's theory did not exclude the role of physics. Like Darwin, D'Arcy Thompson was a wonderful naturalist; unlike Darwin, he was a mathematician. D'Arcy Thompson wanted especially to ensure that the role of physics was not overlooked in explaining biological form. Likewise, the mathematical aspects of his work, which have to do with similarity and transformation, do not of themselves conflict with Darwin's theory of natural selection, either. D'Arcy Thompson's deeper mission was the mathematization of biology. The spirit of mathematics of the day was similarity, and it was reflected in his work on what might be called mathematical biology.

In a lecture he gave in 1911 to the British Association for the Advancement of Science titled "Magnalia Naturae: of The Greater

Problems of Biology," D'Arcy Thompson spoke of a tendency in recent biological work: "the desire to bring to bear upon our science, in greater measure than before, the methods and results of the other sciences, both those that in the hierarchy of knowledge are set above and below, and those that rank alongside our own." He spoke of the unifying influence of physiology, with its focus on the living rather than the dead organism, and its amenability to being treated by the methods of the physical sciences, remarking "Even mathematics has been pressed into the service of the biologist, and the calculus of probabilities is not the only branch of mathematics to which he may usefully appeal." He spoke of the personal appeal that problems about morphology that were related to "mechanical considerations, to mathematical laws, or to physical and chemical processes" held for him. He also laid out reasons supporting the possibility of "so far supporting the observed facts of organic form on mathematical principles, as to bring morphology within or very near to Kant's demand that a true natural science should be justified by its relation to mathematics."

On the first page of the compilation of his ideas into the large compendium *On Growth and Form*, which was published in 1917 and is even now regarded as a masterpiece, he opens the work with a reference to Kant's declaration, as he put it, "that the criterion of true science lay in its relation to mathematics." He goes on to say that:

> As soon as we adventure on the paths of the physicist, we learn to *weigh* and to *measure*, to deal with time and space and mass and their related concepts, and to find more and more our knowledge expressed and our needs satisfied through the concept of number, as in the dreams and visions of Plato and Pythagoras;

D'Arcy Thompson recognized Newton, too, of course, as a prime example of someone whose work had shown the tremendous fruitfulness of mathematizing a class of phenomena. The continuity D'Arcy Thompson saw between his own project of mathematizing biology and Newton, though, was in the use of similarity. "Newton did not shew the cause of the apple falling, but he shewed a similitude ('the more to increase our wonder, with an apple') between the apple and the stars." As in physics, so in the life sciences: "The search for differences or fundamental contrasts between the phenomena of organic and inorganic, of animate and inanimate, things, has occupied many men's minds, while the search for the community of principles or essential similitudes has been pursued by few." He compared the "slow, reluctant extension of physical laws to vital [living] phenomena" with the "slow triumphant demonstration of Tycho Brahe, Copernicus, Galileo and Newton (all in opposition to the Aristotelian cosmogony), that the heavens are formed of like substance with the earth, and that the movements of both are subject to the selfsame laws." D'Arcy Thompson did not go so far as to claim physics and mathematics were comprehensive, nor even to know how much they could explain. He recognized there were limits: ". . . nor do I ask of physics how goodness shines in one man's face, and evil betrays itself in another."

He did look to physics, though, for an explanation of lots of different kinds of behavior in creatures both living and nonliving—including an explanation of heavier-than-air flight. He reiterated the point Boltzmann made "that various capabilities depend in various ways on the linear dimensions," citing Helmholtz's 1873 lecture on similar motions and dirigibles for the reasoning supporting the by-then familiar conclusion that "the work which *can be done* varies with the available weight of muscle, that is to say, with the weight of the bird; but the work which *has to be done* varies with mass and distance; so the larger the bird grows, the greater the disadvantage under which all its work is done." But D'Arcy Thompson goes further and points out that, while this is true for a specific machine or animal form, it is not the whole story. Not all flight is powered by the sources assumed in these analyses. There is also, he says, "gliding flight, in which . . . neither muscular power nor engine power are employed; and we see that the larger birds, vulture, albatross or solan-goose, depend on gliding more and more." This is just one illustration of the fact that many factors other than size are involved in comparing flight capabilities of various birds. These other factors, he says, "vary so much in the complicated action of flight that it is hard indeed to compare one bird with another." In living things, we find that "Nature exhibits so many refinements and 'improvements' in the mechanism required, that a comparison based upon size alone becomes imaginary, and is little worth the making."

What can be said is that, in both how the fish swims and how the bird flies, *streamlining* is important. In properly streamlined wings, "a partial vacuum is formed above the wing and follows it wherever it goes, so long as the stream-lining of the wing and its angle of incidence are suitable, and so long as the bird travels fast enough through the air." Here the kind of reasoning based on the observation Boltzmann had made in his "Models" essay ("that various capabilities depend in various ways on the linear dimensions") is informative: it tells us how the speed required to stay aloft increases with size. D'Arcy Thompson refers to this as a "principle of *necessary speed*," which he describes as "the inevitable relation between the dimensions of a flying object and the minimum velocity at which its flight is stable." That is, "in flight there is a certain necessary speed—a speed (relative to the air) which the bird *must attain* in order to maintain itself aloft, and which *must* increase as its size increases." This principle explains the qualitative differences between large and small birds. Large birds "must fly quickly, or not at all," whereas insects and very small birds such as hummingbirds, are capable of what appears to be "stationary flight," since, for them, "a very slight and scarcely perceptible velocity relatively to the air [is] sufficient for their support and stability."

Thus, a proper understanding of the significance of Boltzmann's observation about capabilities varying with dimension, in conjunction with finer distinctions about what is involved in flight, does permit conclusions based on size: "The ostrich has apparently reached a magnitude, and the moa certainly did so, at which flight by muscular action, according to the normal anatomy of a bird, becomes physiologically impossible. The same reasoning applies to the case of man." But this is not to say that flight is impossible above a certain size—rather, that "gliding and soaring, by which energy is captured from the wind, are modes of flight little needed by the small birds, but more and more essential to the large." So the proper lesson to be drawn from considerations of the dependence of capabilities on size is not the one that had been drawn during the eighteenth century—that humans should not try to fly at all—but rather that humans should learn to *glide*. Thus, he observes, "It was in trying *to glide* that the pioneers of aviation, Cayley, Wenham and Mouillard, Langley, Lilienthal and the Wrights—all careful students of birds—renewed the attempt; and only after the Wrights had learned to glide did they seek to add power to their glider."

What D'Arcy Thompson stressed in all this variety of phenomena was a principle that explained the variety: the *principle of similarity*. Recall that he had mentioned that Newton's insight had involved discerning the similarity underlying the two very different cases of the apple's fall to the earth and the moon's hanging in the sky. In the case of flight, the hummingbird's hanging in the air and the difficulty large birds have in becoming airborne is explained by a principle as well: the *principle of similarity*. Galileo, too, had introduced the principle of similarity by remarking on the *difference* in performance of large and small creatures and machines. Thompson draws out Galileo's point as it pertains to differences in animal form and to engineering design:

> But it was Galileo who, wellnigh three hundred years ago, had first laid down this general principle of similitude; and he did so with the utmost possible clearness, and with a great wealth of illustration drawn from structures living and dead. [cites 1914 translation, p. 130] He said that if we tried building ships, palaces or temples of enormous size, yards, beams and bolts would cease to hold together; nor can Nature grow a tree nor construct an animal beyond a certain size, while retaining the proportions and employing the materials which suffice in the case of a smaller structure. The thing will fall to pieces of its own weight unless we either change its relative proportions, ... or else we must find new material, harder and stronger than was used before. Both processes are familiar to us in Nature and in art, and practical applications, undreamed of by Galileo, meet us at every turn in this modern age of cement and steel.

To "change its relative proportions" is to change an animal's form. Thus, the forms of animals are dependent on, or conditioned on, not only the material properties of the stuff of which they are made, but also of the force of gravity. Form is an effect of scale, but, in turn, "The effect of scale depends not on a thing in itself, but in relation to its whole environment or milieu." If this is so, the form of a land-based animal reflects the strength of the gravitational force. D'Arcy Thompson illustrates the point by asking what things would be like were the gravitational force different:

> Were the force of gravity to be doubled our bipedal form would be a failure, and the majority of terrestrial animals would resemble short-legged saurians, or else serpents. Birds and insects would suffer likewise, though with some compensation in the increased density of the

air. On the other hand, if gravity were halved, we should get a lighter, slenderer, more active type, needing less energy, less heat, less heart, less lungs, less blood. Gravity not only controls the actions but also influences the forms of all save the least of organisms.

For very tiny organisms, the same general principle—that the effect of scale on form is a matter not only of the features of the organism itself, but of its whole environment—applies. However, for motions of such tiny animals, it is not gravity, but surface tension, that tends to be the dominant feature of the environment: "The small insects skating on a pool have their movements controlled and their freedom limited by the surface tension between water and air, and the measure of that tension determines the magnitude which they may attain." There are other constraints on their size due to their form, too. In the respiratory system of insects, "blood does not carry oxygen to the tissues, but innumerable fine tubules or tracheae lead air into the interstices of the body." There are natural limitations on the size of such a system; if they grew too much larger, "a vast complication of tracheal tubules would be necessary, within which friction would increase and fusion be retarded, and which would soon be an inefficient and inappropriate mechanism."

Besides the limitations on size for insect forms, we can, conversely, see the insect's form as constrained by its size: "we find that the form of all very small organisms is independent of gravity, and largely if not mainly due to the force of surface tension." One of D'Arcy Thompson's well-known phrases comes from his point that the form of an object is a "diagram of forces"; the immediate context in which that phrase occurs is as follows:

> The form, then, of any portion of matter, whether it be living or dead, and the changes of form which are apparent in its movements and in its growth, may in all cases alike be described as due to the action of force. In short, the form of an object is a 'diagram of forces,' in this sense, at least, that from it we can judge of or deduce the forces that are acting or have acted upon it:

The point here is the effect on form of the environment, not the importance of forces. D'Arcy Thompson was clear that he was using forces only as a sort of shortcut expression: ". . . force, unlike matter, has no independent objective existence. It is energy in its various forms, known or unknown, that acts upon matter." Here, we recognize his awareness of the view of energeticists (such as Ostwald and Hertz), of the problematic status of forces, and their tendency to replace explanations made in terms of force with explanations in terms of mass and energy. Throughout *On Growth and Form*, we find many explanations of animal behavior and form given in terms of energy available and expended. The reason D'Arcy Thompson used the notion of force in describing form was because form is abstract, rather than material, and he justifies his use of the term "force" as appropriate here *without* reifying force:

> But when we abstract our thoughts from the material to its form, or from the thing moved to its motions, when we deal with the subjective conceptions of form, or movement, or the movements that change of form implies, then Force is the appropriate term for our conception of the causes by which these forms and changes of form are brought about. When we use the term force, we use it, as the physicist always does, for the sake of brevity, using a symbol for the magnitude and direction of an action in reference to the symbol or diagram of a material thing. It is a term as subjective and symbolic as form itself, and so is used appropriately in connection therewith.

He elaborates on the interrelations of magnitude, ratio, and picture: "When we deal with magnitude in relation to the dimensions of space, our diagram plots magnitude in one direction against magnitude in another— length against height, for instance, or against breadth." What we get, he says there, is "what we call a picture or outline, or (more correctly) a 'plane projection' of the object." His emphasis on ratio is striking, and in fact he sums up the whole idea of form in terms of it: "what we call Form is a ratio of magnitudes referred to direction in space." This particular ratio is dimensionless, since it is a ratio of like magnitudes. A length, a height, a breadth, are all measured in dimensions of linear length, whatever units are used. Hence, whatever units are used to measure these magnitudes—inches, feet, millimeters, or centimeters—so long as the same units are used for both the magnitudes in the ratio, the units cancel, and the resulting ratio has no units at all. So any ratio of like magnitudes is dimensionless.

However, Thompson does not restrict the ratios of interest to such ratios. When considering the variation of a length over time, as in studying growth, he points out that the ratio involved there has the dimensions of velocity: "We see that the phenomenon we are studying is a *velocity* (whose 'dimensions' are space/time, or L/T) and this phenomenon we shall speak of, simply, as *rate of growth*." The symbols L and T denote that the dimensions of the magnitudes being measured are length and time, respectively. They do not specify units of measurement, just what kind of measurement is being taken. Constructing a ratio of length to time gives rise to another kind of quantity, and thus we say that the quantity velocity has dimensions of L/T. So ratios of unlike magnitudes give rise to additional kinds of quantities, or kinds of magnitudes. D'Arcy Thompson's graphical representations also used contour-lines, or "isopleths," to represent a third dimension or magnitude on a two-dimensional surface. The contour-lines can show depth, or the third dimension, of a three-dimensional form.

So far, time is represented only insofar as each of these representations represents a form or configuration *at a particular time*. Then, Thompson explains, the outlines of an organism as it changes over time can be set out side by side (or, alternatively, overlaid on each other), and this series represents the organism's gradual change over time. Such a representation—somewhat like a series of comic-strip frames—exhibits both the form of the organism and the growth of the organism. In addition, it shows how an organism's growth and form are interrelated: "it is obvious that the form of an organism is determined by its rate of growth in various directions."

As mentioned earlier, D'Arcy Thompson's goal was to mathematize biology—to treat it the way a physicist treats his subject. He had earlier remarked that "physics is passing through an empirical phase into a phase of pure mathematical reasoning," and certainly the energeticists' emphasis on equations and energy balances was an example of the newer style of mathematical reasoning. Thompson had, however, identified the "old-fashioned empirical physics" as the one "which we endeavour, and are alone able, to apply [when we use physics to interpret and elucidate our biology]." That remark was made in the context of explaining the sense in which it was still appropriate to speak of forces as determining biological form, in spite of his recognition that forces do not exist.

The approaches he mentions in which the variables tracked are velocities seems to reflect at least some features of the Lagrangian approach. Lagrangian mechanics is a reformulation of Newton's formulation of mechanics. In Lagrangian mechanics, energy conservation principles, rather than force balances, are used to solve equations of motion. Lagrangian mechanics by that time was generalized so that the quantities did not even need to be velocities and spatial coordinates; they were instead called "generalized velocities" and "generalized coordinates." It then became possible to express the equations of motion of a system using only variables for generalized velocities and time. The generalized velocities bore the same kind of relationship to generalized coordinates as velocities do to coordinates in classical formulations: they expressed the change in coordinates with respect to time. What was important was that the kinds of quantities that are arguments, or inputs, into the functions that express a system's equations of motion were independent of each other and together characterized the system.

The methods proposed for biology in *On Growth and Form* are easily generalized, so that biology appears as a case of more general principles that apply in physics as well. For, he said, many other things in the world can be seen as cases of the phenomenon of growth, if growth is seen as change in magnitude over time: "since the movement of matter must always involve an element of time, . . . in all cases the *rate of growth* is a phenomenon to be considered." If, as he also said, rate of growth is velocity, what he is saying here is akin to approaches in Lagrangian mechanics, where generalized velocities, rather than forces, are the variables considered important in addition to coordinates of position. We shall see these notions appear elsewhere, such as the use of side-by-side depictions of forms changing over time representing the dynamics of a situation, and the notion of form as consisting of ratios of magnitudes.

Thompson emphasizes the effects of magnitude or size: because different forces are predominant at different scales (gravity at one scale, and surface tension at another), animals on different-sized scales are of very different kinds of form. The mechanical principles that describe these forms and how the growth of these forms is constrained are different at different scales; the different mechanical principles are responsible for the difference in form we observe in animals of very different sizes. In overview, he remarks:

> We found, to begin with, that "scale" had a marked effect on physical phenomena, and that increase or diminution of magnitude might mean a complete change of statical or dynamical equilibrium. In the end we begin to see that there are discontinuities in the scale, defining phases in which different forces predominate and different conditions prevail. . . . [the range of magnitude of life] is wide enough to include three such discrepant conditions as those in which a man, an insect and a bacillus have their being and play their several roles.

He describes what life is like on three different scales, each such "world" smaller than the other:

> Man is ruled by gravitation, and rests on mother earth. A water-beetle finds the surface of a pool a matter of life and death, a perilous entanglement or an indispensable support. In a third world, where the bacillus lives, gravitation is forgotten, and the viscosity of the liquid, the resistance defined by Stokes's law, the molecular shocks of the Brownian movement, doubtless also the electric charges of the ionised medium, make up the physical environment and have their potent and immediate influence on the organism. The predominant factors are no longer those of our scale; we have come to the edge of a world of which we have no experience, and where all our preconceptions must be recast.

However, when Thompson looked for an underlying common principle of which these differing phenomena are illustrative, he found it in the *principle of similitude*, the same one he credited to Galileo in his work on mechanics of materials, and which he says is also recognizable in Newton's explanation of his discovery of the theory of gravitation.

The compilation of D'Arcy Thompson's works into a massive masterwork unified around the theme of growth and form was not published until 1917, but it is based on lots of scientific and engineering work done prior to 1914. He seems to have been especially interested in artificial and natural flight, even citing technical works from the years following the Wright Brothers' 1908 demonstrations in Europe, such as G. H. Bryan's 1911 *Stability in Aviation*, F. W. Lanchester's 1909 *Aerodynamics*, and

George Greenhill's 1912 *The Dynamics of Mechanical Flight*. He cited works more directly relevant to the biological emphasis of the work, too, such as E. H. Hankin's 1913 *Animal Flight*, and many, many scientific papers about insects and other animals.

In *On Growth and Form*, he cited papers from those who had been thinking about aviation well before 1900, too, including Helmholtz's 1873 paper on similar motions and dirigibles, mentioned earlier. Recall that, in that paper, Helmholtz had shown how, even when the differential equations governing the motions of dirigibles could not be solved, one could rewrite the equations in a form such that the coefficients were all dimensionless parameters. Then he showed that any two situations in which these dimensionless parameters were the same would have the same solution. This gives a mathematically sound basis on which one can infer the motions of dirigibles (which were extremely large and unwieldy) from a model or from other observed cases. However, in citing this paper, D'Arcy Thompson does not draw from it anything more general than the kind of reasoning used for a specific case; recall that he was interested in showing that Helmholtz's conclusion held only assuming that the energy keeping a bird aloft came from muscular energy. He did not take issue with Helmholtz's method, only with his assumptions.

Still, it is telling that D'Arcy Thompson does not seem to be interested in the general theory of dimensions and similarity, or even in hydrodynamical similarity, which is rather presciently laid out in the remarkable paper by Helmholtz to which he refers, in which Helmholtz's methodology outpaces both his assumptions and conclusions. It was a paper ahead of its time. Helmholtz had directed his analysis of similar motions to the practical problem of steering air balloons, not gliders or airplanes, though it applied equally to both problems. When research into heavier-than-air flight was pursued in earnest, Helmholtz's paper was resurrected and recognized as containing the basis for all the important dimensionless numbers in hydrodynamics.

This attitude we see in Thompson's treatment of dimensional analysis in *On Growth and Form*—being interested in, even inspired by, the principle of similitude found in Galileo and Newton, yet being content to be led by that interest only so far as necessary to solve a problem at hand—seems to be representative of scientists of that era. Reasoning based on dimensional analysis was used to reach conclusions, and these conclusions were considered basic principles of a general class of problems. Hence, we have various "laws" for specific kinds of situations, such as liquid flow in pipes, boats being towed in canals, streams flowing into lakes, and so on, with a corresponding rule about how a measurement taken on one scale has to be transformed to yield the corresponding value in the situation on another scale. So, for instance, in discussing the speed of aquatic animals, wherein the resistance is provided not by gravitational forces, but by "skin-friction," he reasons:

> Now we have seen that the dimensions of W are l^3 and of R are l^2; and by elementary mechanics $W \propto RV^2$, or $V^2 \propto W/R$. Therefore $V^2 \propto l^3/l^2$, and $V \propto$ square root of l. This is what is known as *Froude's Law*, of the correspondence of speeds—a simple and most elegant instance of "dimensional theory."

He goes on to say that sometimes such questions about the effect of scale are "too complicated to answer in a word."

He points out that, depending on an engine's design, the design work can instead depend on the square, rather than the cube, of linear dimensions, and he mentions a different law in such a case: *Froude's law of steamship comparison*. In a footnote, he cites with approval Lanchester's wry remark that "the great designer was not hampered by a knowledge of the theory of dimensions," which reflects a respect for practical knowledge above this kind of theoretical principle.

Thompson goes on to show that there are subtleties involved that complicate such simplified generalizations. They usually have to do with details of the mechanisms by which different functions are accomplished, so they actually tend to be criticisms of the assumptions used rather than of the methodology of dimensional analysis. One example of such a criticism is the point he had made about analyses based on dimensional reasoning that neglected the importance of gliding flight, which completely reversed the previous conclusion about the possibility of humans achieving heavier-than-air flight. The subtleties, anecdotes, and considerations Thompson brings up are meant to temper looking to any specific derivation using the principle of similitude as the arbiter of effects of scale—that is, to warn against regarding such laws of correspondence as themselves principles of nature. The validity of the particular laws of nature that are consequences of the principle of similitude are very dependent on correct insight into the functions and forces relevant to the behavior of the machine or organism. The principle of similitude, though, is simply a principle about the behavior of forces and functions.

So, despite the reservations expressed, the principle of similitude is called out as the underlying principle of D'Arcy Thompson's book, and with appropriate justification:

> In short, it often happens that of the forces in action in a system some vary as one power and some as another, of the masses, distances or other magnitudes involved; the "dimensions" remain the same in our equations of equilibrium, but the relative values alter with the scale. This is known as the "Principle of Similitude," or of dynamical similarity, and it and its consequences are of great importance.

Thompson's conclusion here about the significance of the principle of similarity, or dynamical similarity, follows his more specific observations that:

> A common effect of scale is due to the fact that, of the physical forces, some act either directly at the surface of a body, or otherwise in proportion to its surface or area; while others, and above all gravity, act on all particles, internal and external alike, and exert a force which is proportional to the mass, and so usually to the volume of the body.

Thus, the principle of similitude is a very general principle of science—of any science using measurement, in fact. The criticisms and qualifications D'Arcy Thompson raised were directed at *inappropriate uses* of the specific "laws" and "laws of correspondence" derived from considerations of similarity and theory of dimensions, not at the principle of similitude itself. Usually, the principle of similarity states conditions of similarity in terms of the constancy of a certain dimensionless parameter, and hence states when similar motions are expected, within a certain "scale" or "world." However, there are often discontinuities in behavior, such as when a flow transitions from smooth to turbulent, or as a substance transitions from a liquid to a gas. D'Arcy Thompson seems to cite the principle of similarity as an explanation of both the similarities that can be drawn within a particular scale and the discontinuities that exist between scales.

In fact, the way Thompson has stated the principle, it explains not only similarities and differences in behavior of animals and artificial machines within the same one of the three "worlds" he describes, but it also accounts for the discontinuities and very different kinds of forms and forces encountered between those three "worlds."

Physical similarity, dynamical similarity, and principles of similarity or similitude became more prominent topics in 1913 and early 1914. For one thing, an English translation of Galileo's *Dialogue Concerning Two New Sciences* was published in February 1914, after being practically unobtainable. It was bound to be of great interest to anyone interested in the history or philosophy of science. According to the Translator's Preface by Henry Crew and Alfonso De Salvio, copies of the previous translation of *Two New Sciences* into English, done in 1730 by Thomas Weston, had become "scarce and expensive" by 1914. An even earlier English translation had "issued from the English press in 1665." But fate intervened and, they said, "It is supposed that most of the copies were destroyed in the great London fire which occurred in the year following. . . . even [the copy] belonging to the British Museum is an imperfect one." The drama of the long struggle for this work to become available in English to twentieth-century readers made it a publishing event. At least, that is how the translators saw it:

> For more than a century English speaking students have been placed in the anomalous position of hearing Galileo constantly referred to as the founder of modern physical science, without having any chance to read, in their own language, what Galileo himself has to say. Archimedes has been made available by Heath; Huygens' *Light* has been turned into English by Thompson, while Motte has put the *Principia* of Newton back into the language in which it was conceived. To render the Physics of Galileo also accessible to English and American students is the purpose of the following translation.

Crew lectured on the topic of the things to be learned from Galileo in 1913, and he mentioned the "theory of dimensions" in passing, even before the English translation of *Two New Sciences* was published.

According to the recent study of Wittgenstein's books left with Russell (as mentioned earlier), which were books Wittgenstein owned during from 1905 to 1913, Wittgenstein owned one of the "scarce and expensive" editions of Galileo's *Discourses Concerning Two New Sciences* published in London in 1730. It is the only book by Galileo in the collection of Wittgenstein's books surveyed. Oystein Hide, the author of the study, remarks that "Wittgenstein later drew comparisons between his philosophical activity and the work of, for example, Galileo within his scientific field."

The collection of books owned by Wittgenstein that Hide describes does not contain many other books of this sort, with two striking exceptions: a six-volume set of facsimiles of notebooks of "Leonardo De Vinci" published in Paris in 1891, and a copy of *Principia Philosophiae* by "Isaco Newtono" published in 1728. Galileo and Newton stand out as the early scientists who wrote on similarity and similitude. We saw that D'Arcy Thompson cited Newton's use of similitude in discovering the law of gravitation, and that Boltzmann cited Newton in his encyclopedia article on models just after mentioning models of flying machines, in mentioning "The theory, initiated by Sir Isaac Newton, of the dependence of various effects on the linear dimensions." And we shall see that the physicist Heike Kamerlingh Onnes mentions Newton's use of mechanical similitude in his 1913 Nobel lecture. Leonardo da Vinci is known not only for his work on flying machines, including parachutes and helicopters, but for his work in hydrodynamics and his explicit discussions and illustrations of the use of proportion in art as well as science. His notebooks used both his artistic capabilities and his mechanical genius. His sketch of the human body with geometric proportions overlaid on it is well-known, but he investigated many topics, and his work in hydrodynamics was striking. His sketches of fluid flows, especially turbulent flows, show a remarkable talent for observation of the phenomena that were being investigated with such intensity in the years prior to 1914. He is also known for his use of proportion in art as well as in science. It's possible that the theft of the Mona Lisa from the Louvre in 1911 and its recovery in 1913 put Leonardo's works in the public eye during this time. At any rate, the fact that among the books Wittgenstein owned prior to 1914, we find expensive copies of works by Newton and Galileo in which they discuss similitude may reflect interests stimulated by the resurgence of interest in their work on similitude in England (and perhaps elsewhere) in the years preceding 1914. The interest in Leonardo's notebooks can be accounted for by the aeronautical work in his notebooks alone, especially his invention of a helicopter, but it also fits with an interest in ratio and proportion.

The publishing event of a new English translation of Galileo's *Two New Sciences* in 1914, which was being written about and discussed in 1913 prior to its publication, is significant to our story because of the prominence of the discussion of similarity it contains and the accessible and memorable way the principle is explained and its consequences depicted. One of the translators, Henry Crew, was a physicist, a professor of physics at Northwestern University, and had studied with Hermann von Helmholtz in Berlin in 1883–1884; this would have been after Helmholtz had already developed the criteria for similar motions, and perhaps it accounts for Crew's interest in translating this particular work of Galileo's. In a nutshell, the dialogue begins with a conversant using a limited notion of similarity (geometric similarity), puzzling over the invalidity of the consequences one can draw from it, and then proceeds to a discussion in which the same conversant comes to use a more general notion of similarity, from which the conclusions and experimental predictions drawn are in fact valid. There are not, as in D'Arcy Thompson's treatment of the principle, examples about flight, but there is more emphasis on correct, rather than incorrect, uses of the principle.

Galileo's dialogue begins with Salviati, usually taken to be the voice of Galileo, recounting numerous examples of a large structure that has the same proportions and ratios as a smaller structure but that is not proportionately strong. In these opening pages of the dialogue, the wise and seasoned Salviati explains to the earnest but puzzled Sagredo that "if a piece of scantling [corrente] will carry the weight of ten similar to itself, a beam [trave] having the same proportions will not be able to support ten similar beams." The phenomenon of the effect of size on the function of machines of similar design holds among natural as well as artificial forms, Salviati explains: "just as smaller animals are proportionately stronger and more robust than the larger, so also smaller plants are able to stand up better than larger." Perhaps the most well-known of Salviati's illustrations is about giants:

> . . . an oak two hundred cubits high would not be able to sustain its own branches if they were distributed as in a tree of ordinary size; and [. . .] nature cannot produce a horse as large as twenty ordinary

horses or a giant ten times taller than an ordinary man unless by miracle or by greatly altering the proportions of his limbs and especially his bones, which would have to be considerably enlarged over the ordinary. Likewise the current belief that, in the case of artificial machines the very large and the small are equally feasible and lasting is a manifest error.

It is this point about scale we saw reflected throughout D'Arcy Thompson's work, applied especially to the case of flight. Yet it was well known in 1914 that engineers used scale models. The question was whether a certain use was valid, and the problem was that determining this often seemed a matter of engineering knowledge and skill, not a matter of pure science. Recall that there was some ambivalence in D'Arcy Thompson's discussion; he criticized a number of analyses based on dimensional considerations, although he is clear that his reservations were not about the validity of the principle of similitude itself. He credited Galileo as the first to articulate the principle of similitude, or dynamical similarity.

Although Galileo's work opens with the wise participant in the dialogue reminding the others of the reasons for the lack of giant versions of naturally occurring life-forms, it soon proceeds to a case of a valid use of a small (artificial) machine to infer the behavior of a large (artificial) machine. As in Helmholtz's reasoning in his paper on similar motions and dirigibles, the basis for similarity is found not in mere geometric similarity, but, more deeply, in dimensional considerations drawn from an equation of motion. At a later point in the dialogue, Sagredo makes use of Salviati's statement that the "times of vibration" (period of oscillation) of bodies suspended by threads of different lengths "bear to each other the same proportion as the square roots of the lengths of the thread; or one might say the lengths are to each other as the squares of the times." From this, Sagredo uses one physical pendulum to infer the length of another physical pendulum:

> Then, if I understand you correctly, I can easily measure the length of a string whose upper end is attached at any height whatever even if this end were invisible and I could see only the lower extremity. For if I attach to the lower end of this string a rather heavy weight and give

it a to-and-fro motion, and if I ask a friend to count a number of its vibrations, while I, during the same time-interval, count the number of vibrations of a pendulum which is exactly one cubit in length, then knowing the number of vibrations which each pendulum makes in the given interval of time one can determine the length of the string. Suppose, for example, that my friend counts 20 vibrations of the long cord during the same time in which I count 240 of my string which is one cubit in length; taking the squares of the two numbers, 20 and 240, namely 400 and 57600, then, I say, the long string contains 57600 units of such length that my pendulum will contain 400 of them; and since the length of my string is one cubit, I shall divide 57600 by 400 and thus obtain 144. Accordingly I shall call the length of the string 144 cubits.

The basis on which Sagredo infers the length of the larger from the smaller is a fundamental relationship—the relationship between length and period that describes the behavior of any pendulum, which can be expressed in terms of the constancy of the value of a certain ratio containing them. What he derives from it is a *law of correspondence* telling him how to find the corresponding length in the large pendulum from the length of the small. Salviati (the voice of Galileo) responds approvingly to his claim that this method will yield the length of the string: "Nor will you miss it by as much as a hand's breadth, especially if you observe a large number of vibrations." Henry Crew, the physicist who co-translated the work, thought the emphasis on experiment in *Two New Sciences* was one of the important contributions of Galileo's work.

This work of Galileo's, which is generally credited with giving not only the first, but probably the best ever, exposition of physical similarity, appeared just as scientists from Britain's National Physical Laboratory (NPL) presented a compendium of their own about similarity. The paper, "Similarity of Motion in Relation to the Surface Friction of Fluids," by T. E. Stanton and J. R. Pannell, was submitted to the Royal Society of London in December 1913 and was read to the Society in January of 1914. Stanton was superintendent of NPL's Engineering Department and was interested in the possibilities of using small-scale models in wind tunnels for engineering research. Except for the work of Osborne Reynolds, experimental study of similar motions in fluids was, according to the authors, only done since 1909:

Apart from the researches on similarity of motion of fluids, which have been in progress in the Aeronautical Department of the National Physical laboratory during the last four years, the only previous experimental investigation on the subject, as far as the authors are aware, has been that of Osborne Reynolds . . .

Osborne Reynolds was the celebrated professor of engineering at Manchester who had retired a few years prior to Wittgenstein's arrival and enrollment there as an engineering student. What Stanton and Pannell cite as Reynolds' major discoveries were that there was a critical point at which fluid flow suddenly changed from "lamellar motion" to "eddying motion" (today we would say "from laminar flow to turbulent flow"), that the critical velocity was directly proportional to the kinematical viscosity of the water and inversely proportional to the diameter of the tube, and that for geometrically similar tubes, the dimensionless product:

(critical velocity) × (diameter) / (kinematic viscosity of water)

was constant. They also noted that, no matter what the conditions of flow, whether above or below the critical velocity, whenever the values of the dimensionless product:

(velocity) × (diameter) / (kinematic viscosity of water), or $\dfrac{vd}{v}$

were the same, so were the corresponding values of another dimensionless parameter:

(density) × (diameter)^3 / (coefficient of viscosity of water)^2
  × (rate of fall of pressure along the length of the pipe)

As is often the case, there was a complication: experiment showed that the surface roughness of the pipe wall (or of whatever surface forms the flow boundary) needed to be taken into account as well. This is a matter of geometry on a much smaller scale making a difference. However, the overall approach of the use of dimensionless parameters to establish similar situations was still seen to be valid, as their experiments illustrated:

From the foregoing it appears that similarity of motion in fluids at constant values of the variable v d / v will exist, provided the surfaces relative to which the fluids move are geometrically similar, which similarity, as Lord RAYLEIGH has pointed out, must extend to those irregularities in the surfaces which constitute roughness. In view of the practical value of the ability to apply this principle to the prediction of the resistance of aircraft from experiments on models, experimental investigation of the conditions under which similar motions can be produced under practical conditions becomes of considerable importance, ... By the use of colouring matter to reveal the eddy systems at the back of similar inclined plates in streams of air and water, photographs of the systems existing in the two fluids when the value of v d / v was the same for each, have been obtained, and their comparison has revealed a remarkable similarity in the motions. [ref: *Report of Advisory Committee for Aeronautics*, 1911–1912, p. 97]

The authors here refer to the dimensionless parameter v d / v as a variable. This variable is a product of several measurable quantities and is a dimensionless parameter. That is, the units all cancel out, and its value is independent of the choice of units of measurement. To see this, we can talk about the dimensions of each contributor to the dimensionless parameter, also known as a dimensionless product: the dimensions of v are length divided by time; the dimensions of d are length, and the dimensions of v are (length times length) divided by time. What Stanton and Pannell meant in referring to it as a variable was that their equation for the resistance R includes a function of this dimensionless parameter:

resistance R = (density) × (velocity)^2 × (some function of v d / v)

Or, as they put it, R = ρ v^2 F (v d / v), where F (v d / v) indicates some unspecified function of v d / v. Hence, v d / v is a variable in the sense that the relation for resistance includes an unspecified function of v d / v. It is also a variable in a more practical sense: it can be physically manipulated.

Stanton and Pannell present this relation as a consequence of the Principle of Dynamical Similarity, in conjunction with assumptions about what "the resistance of bodies immersed in fluids moving relatively to them" depends on. Evidently, it was Rayleigh who suggested the generalization; they cite Rayleigh's contribution in the *Report of the Advisory Committee for Aeronautics*, 1909–1910, p. 38. Rayleigh had there spoken of the possibility of taking a more general approach than current researchers were taking in applying "the principle of dynamical similarity." He explained his "more general" approach as follows:

. . . We will commence by supposing the plane of the plate perpendicular to the stream and inquire as to the dependence of the forces upon the linear dimension (l) of the plate and upon the density (ρ), velocity (v) and kinematic viscosity (ν) of the fluid. Geometrical similarity is presupposed, and until the necessity is disproved it must be assumed to extend to the thickness of the plate as well as to the irregularities of surface which constitute roughness.

If the above-mentioned quantities suffice to determine the effects, the expression for the mean force per unit area normal to the plate (P), analogous to a pressure, is

$$P = r\,v2\,.\,f\,(n\,/\,v\,l) \dots\dots\dots\dots\dots\dots\dots\dots\dots\dots\dots\dots(A)$$

where f is an arbitrary function of the one variable n / v l.

It is for experiment to determine the form of this function, or in the alternative to show that the facts cannot be represented at all by an equation of form (A).

Rayleigh does not here say exactly how the relation (A) is obtained from the assumptions he lists, which was not uncommon at that time when invoking the principle of similarity. Referring to his other papers, and papers by others who invoked it, such as Reynolds and Helmholtz, it is fairly clear that invoking the principle of similarity meant using dimensional considerations, sometimes using the specific fact that a scientific equation required consistent units.

Stanton and Pannell present the results obtained at the National Physical Laboratory in the paper. It is interesting to note that the results are presented in graphs where one of the variables plotted is the term $R / \rho\,v^2$, which is just another expression for the unspecified function, and is dimensionless. What this implies is that the laboratory experiments are not conceived of in terms of the values of individual measurable quantities such as velocity but are classified in terms of the value of a dimensionless parameter.

Rayleigh, too, published a kind of survey paper in early 1914. Here he actively campaigned for wider appreciation and use of the principle, which he credited Stokes with having "laid down in all its completeness." Rayleigh wrote: "There is a general law, called the law of dynamical similarity, which is often of great service. In the past this law has been unaccountably neglected, and not only in the present field. It allows us to infer what will happen upon one scale of operations from what has been observed at another." He then discussed the example of a sphere moving uniformly through air, remarking that, if the kinematic viscosity can be assumed to be the same across the cases considered:

> When a solid sphere moves uniformly through air, the character of the motion of the fluid round it may depend upon the size of the sphere and upon the velocity with which it travels. But we may infer that the motions remain *similar*, if only the product of diameter and velocity be given. Thus if we know the motion for a particular diameter and velocity of the sphere, we can infer what it will be when the velocity is halved and the diameter doubled. The fluid velocities also will everywhere be halved at the *corresponding* places.

So we see one use of the principle is to be able to use one observation or experiment as representative of a whole class of actual cases: all the other cases to which it is similar, even though the cases may have very different values of measurable quantities such as velocity. The important fact of the situation is the dimensionless parameter just mentioned: "It appears that similar motions may take place provided a certain condition be satisfied, viz. that the product of the linear dimension and the velocity, divided by the kinematic viscosity of the fluid, remain unchanged."

The important feature of the situation is the value of this dimensionless parameter; that would mean that, even in cases of a *different fluid*, so long as this dimensionless product is the same, the motions will be similar! Can that be right? Yes, and Rayleigh points out that it is a particularly useful application of the principle: "If we know what happens on a certain scale and at a certain velocity in *water*, we can infer what will happen in *air* on any other scale, provided the velocity is chosen suitably."

There is a qualification he adds here, one that the reader familiar with the qualifications needed since the advent of special relativity (that relativistic phenomena do not appear only so long as the velocities involved are much smaller than the velocity of light) might see as an analogue for sound: "It is assumed here that the compressibility of the air does not come into account, an assumption which is admissible so long as the velocities are small in comparison with that of sound." Neglecting the compressibility of air in certain velocity ranges and not others is another illustration of the discontinuity of scale D'Arcy Thompson emphasized. It does not mean that the method of similarity requires neglecting the compressibility of air, just that in the low-velocity range, establishing similarity of motions between two different situations does not require accounting for it.

However, Rayleigh also pointed out that, in contrast to the compressibility of air, it appeared that viscosity *was* important in many cases where it was so small that it seemed improbable that it should matter. When viscosities were low, as in water, one would not expect that the actual value of viscosity would be a significant factor in water's qualitative behavior. Osborne Reynolds' results on fluid flow in pipes had shown that it is; Reynolds began to suspect that viscosity was important even in water when he observed unexpected changes in fluid flow as the temperature was varied. Since viscosity varies with temperature, he investigated the effect of viscosity and found that it was indeed important for fluid flow through pipes, even for nonviscous fluids such as water. He also investigated cases where viscosity was the "leading consideration," as Rayleigh put it, in remarking that "It appears that in the extreme cases, when viscosity can be neglected and again when it is paramount, we are able to give a pretty good account of what passes. It is in the intermediate region, where both inertia and viscosity are of influence, that the difficulty is greatest."

These unexpected experimental results showing that viscosity needed to be taken into account in these intermediate regions were as unwelcome as they were unexpected, since there was already on hand a well-developed theory of hydrodynamics that neglected viscosity—one that lent itself to closed-form mathematical solutions of the equations. This discipline, hydrodynamics, had produced many beautiful mathematical theorems and solutions. If fluid friction were included, though, the mathematical equations became intractable, as Helmholtz had remarked in his papers on hydrodynamics. Helmholtz had lain out and responded to this problem in his 1873 paper on similar motions of air balloons; his response had been to show how to use the intractable equations to find similarity conditions—the relevant dimensionless parameters for dynamical similarity—to permit inferring the behavior of a large balloon from experiments on smaller ones. This was an experimental alternative to the impossible task of finding a solution of the intractable hydrodynamical equations that accounted for fluid friction, or viscosity. As the authors of an historical survey put it: "Helmholtz, in fact, presented to the Berlin Academy of Sciences in 1873 a dimensional analysis of the equations of fluid motion that already encompassed what we know today as the Froude, Reynolds, and Mach criteria for model-prototype similarity." This is what Rayleigh was referring to when he added that, even in the intermediate region, where the difficulty is the greatest, "we are not wholly without guidance," lamenting the neglect of "the law of dynamical similarity."

One might think that, by 1914, when the use of wind tunnels had become recognized as essential to practical aeronautical research, this principle would have become accepted and would no longer be in question, at least among aeronautical researchers. But if Rayleigh's estimation of the state of the profession is correct, even as late as 1914 this wasn't so! He writes that "although the principle of similarity is well established on the theoretical side and has met with some confirmation in experiment, there has been much hesitation in applying it, . . ." He especially mentions problems in its acceptance in aeronautics due to skepticism that viscosity, which is extremely small in air, should be considered an important parameter: "In order to remove these doubts it is very desirable to experiment with different viscosities, but this is not easy to do on a moderately large scale, as in the wind channels used for aeronautical purposes."

Rayleigh tries to persuade the reader of the significance of the effects of viscosity on the velocity of fluid flow by relating some experiments he performed with a cleverly designed apparatus in his laboratory. The apparatus consisted of two bottles containing fluid at different heights, connected by a tube with a constriction, through which fluid flowed due to the difference in "head," or height of fluid, in the two bottles. The tube with the constriction contained fittings that allow measurement of pressure head at the constriction, and on either side of it. To investigate the effects of viscosity, Rayleigh varied the temperature of the fluid, which changes the fluid viscosity, and he observed how the velocity of the fluid flowing between the two bottles was affected. The kind of relationship he establishes and uses is of the form Galileo employed in reasoning from one pendulum to another. In other words, he worked in terms of ratios (ratios of velocities, ratios of viscosities, ratios of heads), and he employed the fact that some ratios are the square root of others. He took the experimental results he reported in this 1914 paper to conclusively settle the question of the relevance of viscosity to fluid motions.

Thus, in early 1914, there was first a report from Britain's National Physical Laboratory on the experimental work performed there recently on similarity of fluid motions, presented in January; then, in March, Rayleigh's paper on Fluid Motions appeared. Rayleigh's paper explained the need fulfilled by such experimental work, which employed the principle of dynamical similarity and reported his own experiments meant to eradicate any remaining skepticism about the principle. In these works, there is a move toward generalization of the specific relationships then in use for specific applied problems in hydraulics. The work at the National Physical Laboratory was aimed at carrying out a "systematic series of experiments" for "the purpose of establishing a general relation which would be applicable to all fluids and conditions of flow." The dual purpose of both establishing the relation and exploring its limits is seen in how Stanton and Pannell stated their purpose in the paper submitted in December 1913:

> The object of the present paper is to furnish evidence confirming the existence, under certain conditions, of the similarity in motions of fluids of widely differing viscosities and densities which has been predicted, and further by extending the observations through a range in velocity of flow which has not hitherto been attempted to investigate the limits of accuracy of the generally accepted formulae used in calculations of surface friction.

The immediate motivation for performing experiments to furnish this evidence was practical, though there was no hope of producing an equation from which predictions could be directly generated. Recall that the general relation the paper proposed contained an *undetermined* function. So the goal was to find out how to find situations similar to the one you needed to know about that could be carried out in the laboratory and to use the correspondence between them to determine what you wanted to know. It was the question of similar motions that was of special interest in the investigation: "experimental investigation of the conditions under which similar motions can be produced under practical conditions becomes of considerable importance" because of "the practical value of the ability to apply this principle to the prediction of the resistance of aircraft from experiments on models." To be brief, what they were really interested in was providing a methodology for experimental engineering scale models in hydrodynamics and aerodynamics. Somehow the key lay not in a fully detailed equation, but in an equation that enabled them to determine when certain situations were similar and told them how to determine corresponding motions in one situation from observations made on a similar situation.

Adding to this swirl of ideas gathering in late 1913 and early 1914 was the attention being paid to a wonderful experimental result based on similarity and correspondence in the area of thermodynamics: the successful liquification of helium. The same month that Stanton and Pannell submitted their paper from the National Physical Laboratory, December 1913, Heike Kamerlingh Onnes delivered his Nobel lecture. Again, the ideas were not brand-new in 1913, but the worldwide public attention being paid to them was. Onnes had published a paper in 1881 called "General Theory of Liquids," in which he argued that van der Waals' "Law of Corresponding States," which had just been published the prior year, could be derived from scaling arguments, in conjunction with assumptions about how molecules behaved. Van der Waals was impressed with the paper, and a long friendship between the two ensued. Van der Waals won the Nobel Prize in 1910 for "The equation of state for gases and liquids," and he mentioned in his lecture that the "law of corresponding states" had become "universally known," though he said nothing more about it there. The Nobel Prize awarded to Onnes in 1913 brought wider attention to Onnes's account of the foundations of the law of corresponding states—though you might not guess it from the title of Onnes's award, given for "Investigations into the properties of substances at low temperatures, which have led, amongst other things, to the preparation of liquid helium."

Onnes's lecture highlighted the connection between his investigations into properties of substances at low temperatures and similarity principles:

> ... [F]rom the very beginning ... I allowed myself to be led by Van der Waals' theories, particularly by the law of corresponding states which at that time had just been deduced by Van der Waals.
>
> This law had a particular attraction for me because I thought to find the basis for it in the stationary mechanical similarity of substances and from this point of view the study of deviations in substances of simple chemical structure with low critical temperatures seemed particularly important.

What's special about low temperatures, of course, is that, according to the kinetic theory of gases on which van der Waals' equation of state was based, there would be much less molecular motion. Onnes's approach in looking for the foundation of the law of corresponding states has a slightly different emphasis than the kinetic theory of gases. Boyle's Law (often called the ideal gas law) and van der Waals' equation were based on investigating the relationship between the microscale (the molecular level) and the macroscale (the properties of the substance, such as temperature and density). But Onnes was instead looking at the foundation for the similarity of states. Like van der Waals, he looked to mechanics and physics for governing principles, but Onnes pointed out that it was also useful to look at principles of similarity. At low enough temperatures, where motion of the molecules was not the predominant factor, the relevant principles of similarity would be principles of static mechanical similarity, as opposed to dynamical similarity.

As had happened in Osborne Reynolds' work in hydrodynamics, a criterion for similarity had arisen out of investigations into the transition from one regime to another. For Reynolds, it was the critical point at which fluid flow underwent a transition from laminar to turbulent flow (or, in his terminology, from "lamellar" to "eddying" flow) that led to the identification of the dimensionless parameter that later became known as Reynolds Number. The Reynolds Number is in a way a criterion of similarity: fluid systems with the same Reynolds Numbers will be in the same flow regime, regardless of the fluid. So it was with thermodynamics: the critical point at which a substance undergoes a transition from the gaseous to liquid state led to the identification of a criterion of similarity of states that held for all substances.

Van der Waals was very interested in the continuity of states, particularly the continuity between the gas and liquid states. Although his prize was awarded for his work on "The equation of state for gases and liquids," the doctoral dissertation in which he presented the equation had been titled "On the continuity of the gas and liquid state," and he emphasized the important role that the idea of continuity had played in developing the equation:

> . . . I conceived the idea that there is no essential difference between the gaseous and the liquid state of matter—that the factors which, apart from the motion of the molecules, act to determine the

pressure must be regarded as quantitatively different when the density changes and perhaps also when the temperature changes, but that they must be the very factors which exercise their influence throughout. And so the idea of continuity occurred to me.

Van der Waals thought that the force of attraction between molecules would be a big part of the story about the continuity of the gas and liquid states, so he wanted to revise the equation to account for it. He added two more variables to Boyle's Law, which he called "a" and "b" and which are characteristic of a particular substance. Instead of Boyle's ideal gas law—(pressure) × (volume) = (gas constant) × (temperature)—van der Waals' equation can, as Johanna Levelt Sengers points out, be written as a cubic equation—that is, as a third-degree polynomial in volume, with coefficients depending on temperature and pressure. There are many variations on the equation he proposed, but the specific form of the equation and the details about the various ways in which it was revised are not important to us here. What is significant about it for us here is that van der Waals could use the equation to explicitly solve for the values of volume, temperature, and pressure at the critical point. The solution of the equation at the critical point is thus for the trio and is in terms of the parameters a and b introduced by van der Waals:

$$P_c = a / (27\, b^2)$$
$$V_c = 3b$$
$$RT_c = 8a / (27\, b)$$

The specific expressions are not important to our story here; the important thing to notice about them is that the critical values of pressure, volume, and temperature can be expressed in terms of the parameters a and b that van der Waals introduced, and that a and b are characteristic of a substance. Once these are in hand, the next conceptual step is to use these expressions to *eliminate* the constants a, b, and R from his equation of state. He does this by using the critical values of these measurable quantities as units of measurement. Or, as we might say today, he normalizes pressure, volume, and temperature using their critical values, just as Mach number is a ratio of the velocity of something (such as a bullet) in a medium (such as air) to the celerity, or velocity of sound in the medium. It is only because their critical values are given as functions of the constants a and b that his reduction is possible. Thus, reduced pressure, denoted by P*, is the pressure of the gas or liquid divided by P_c, and V* and T* are similarly defined. This yields an equation of state in which neither a, b, nor R appears. Levelt Sengers writes, "This is a truly remarkable result." The equation:

> . . . is *universal*: all characteristics of individual fluids have disappeared from it or, rather, have been hidden in the reduction factors. The reduced pressures of two fluids are the same if the fluids are in *corresponding states*, that is, at the same reduced volume and pressure.

In his presentations to the Academy, van der Waals (1880a, b) deduces straightforwardly that in reduced coordinates, the vapor pressure curve and the coexistence curve must be the same ("fall on top of each other") for all fluids.

The principle of corresponding states allowed scientists to produce curves representative of all substances from experiments on a particular substance:

> The principle of corresponding states . . . frees the scientist from the particular constraints of the van der Waals equation. The properties of a fluid can now be predicted if only its critical parameters are known, simply from correspondence with the properties of a well characterized reference fluid. Alternatively, unknown critical properties of a fluid can be predicted if its properties are known in a region not necessarily close to criticality, based on the behavior of the reference fluid.

If we reflect on how this method of prediction works, we see that the same could be said of Reynolds' discoveries about the critical point of transition from laminar to turbulent flow. Although people today tend to think in terms of a critical Reynolds Number—that is, the value the dimensionless parameter has at the point of transition—in their 1914 paper, Stanton and Pannell put Reynolds' discovery in terms of a critical velocity. By using v_c as an abbreviation for critical velocity, this can be conveniently abbreviated by saying, as they put it, that "Reynolds's discovery was that for geometrically similar tubes v_c d / ν was constant," where d denotes a diameter or other chosen distance in the situation, and ν denotes viscosity. This is stating Reynold's discovery as the kind of constraint provided in thermodynamics by the statement that the dimensionless parameter $(P_c \times V_c) / (R \times T_c)$ is constant. Don't be misled by the fact that these look like the kind of equations you are familiar with using to plug in values and predict the one chosen as the unknown. As in thermodynamics, so in hydrodynamics: their value lies not in using the equation directly, but in telling how experimental curves for one substance can be used to predict the behavior of another. The curves for fluid flow were meant to apply to any fluid—hence Rayleigh's comments that experiments on oil as a reference fluid could be used to predict the critical velocity of another fluid from its properties in a noncritical region—just as the law of corresponding states allows one to make predictions about the critical points of other substances from experiments performed on a reference substance.

Onnes used this insight about corresponding states to set up an experimental apparatus to liquefy helium, which has an extremely low critical temperature. What is so exciting about his story is that he had to rely on the law of corresponding states to estimate the critical temperature so that he would know where to look—that is, so that he would know what conditions to create in order for helium to liquefy. What is especially relevant to our story is that he did more than just use van der Waals' law of corresponding states. He also gave a foundation for it that was independent of the exact form of van der Waals' equation and did not depend on results in statistical mechanics. Instead, he used *mechanical* similarity:

> Kamerlingh Onnes's (1881) purpose is to demonstrate that the principle of corresponding states can be derived on the basis of what he calls the principle of similarity of motion, which he ascribes to Newton. He assumes, with Van der Waals, that the molecules are elastic bodies of constant size, which are subjected to attractive forces only when in the boundary layer near a wall, since the attractive forces in the interior of the volume are assumed to balance each other. . . . He realizes this can be valid only if there is a large number of molecules within the range of attraction. . . . [Onnes] considered a state in which N molecules occupy a volume v, and all have the same speed u (no Maxwellian distribution!). The problem is to express the external pressure p, required to keep the system of moving particles in balance, as a function of the five parameters. He solves this problem by deriving a set of scaling relations for M, A, v, u and p, which pertain if the units of length, mass, and time are changed.

The "scaling relations" Onnes developed are another way of bringing in dimensional considerations, or the "theory of dimensions" that we saw earlier as key to D'Arcy Wentworth Thompson's work in biology on the importance of size, to Galileo's arguments about similarity in mechanics of materials, to Helmholtz's paper on similar motions and dirigibles, to the work by Reynolds, Stanton, and Pannell on similar motions in fluids, and to Rayleigh's crusade for a proper appreciation and more widespread use of the principle of dynamical similarity. As Sengers notes, scaling relations are supposed to hold no matter what units are used for the measurable quantities involved. Onnes provides a criterion for corresponding states based on these scaling relations, along with assumptions about what the molecular-sized objects are like. Sengers remarks:

> Two fluids are in corresponding states if, by proper scaling of length, time and mass for each fluid, they can be brought into the same "state of motion." It is not clearly stated what he means by this, but he must have had in mind an exact mapping of the molecular motion in one system onto that of another system if the systems are in corresponding states.

Then she gives her own suggestion of what it means to be in the same "state of motion":

> In modern terms: suppose a movie is made of the molecular motions in one fluid. Then, after setting the initial positions and speed of the molecules, choosing the temperature and volume of a second fluid appropriately, and adjusting the film speed, a movie of the molecular motion in a second fluid can be made to be an exact replica of that in the first fluid.

Shortly after Onnes's Nobel lecture, Richard Chase Tolman, a physicist at the California Institute of Technology, published a paper titled "The Principle of Similitude" in *Physical Review*, a major physics journal in the U.S. What it suggested sounded a lot like the idea of being able to make movies of one situation that were replicas of movies of other situations except for the film speed. Tolman's paper proposed the following:

> The fundamental entities out of which the physical universe is constructed are of such a nature that from them a miniature universe could be constructed exactly similar in every respect to the present universe.

Tolman then proceeded to show that he could derive a variety of laws, including the ideal gas law, from the principle of similitude he proposed. He proceeded in much the same way as Onnes had done in showing that the principle of corresponding states was a consequence of mechanical similarity. Tolman first developed scaling laws, laying out a transformation rule for how each quantity from a short list he had constructed—length, time, velocity, acceleration, electrical charge, and mass—should be scaled from the present universe to the miniature universe. He laid out and justified a transformation rule for each quantity individually, but some quantities were dependent on others (the transformation equation for velocity is a consequence of the equations for length and time). His justifications seem to be based on the criterion that the two universes would be observationally equivalent from the standpoint of an observer located in one of them:

> . . . let us consider two observers, O and O', provided with instruments for making physical measurements. O is provided with ordinary meter sticks, clocks and other measuring apparatus of the kind and size which we now possess, and makes measurements in our present physical universe. O', however, is provided with a shorter meter stick, and correspondingly altered clocks and other apparatus so that he could make measurements in the miniature universe of which we have spoken, and in accordance with our postulate obtain exactly the same numerical results in all his experiments as does O in the analogous measurements made in the real universe.

Examples are that if O measures a length to be l, O' measures it to be xl, where O' has a meter stick that is shorter than O's and x is a number less than 1. In obtaining the transformation equation for time, however, Tolman appeals to the physical fact that "the velocity of light in free space must measure the same for O and O'," and he concludes that if O measures t, O' will measure xt. So the equations for length and time are l' = xl and t' = xt. In obtaining the transformation equation for mass, he appeals to Coulomb's law as setting a constraint that must be satisfied, in conjunction with the requirement that O and O' measure the same charge, to obtain m' = m / x. Once the equations for the "fundamental magnitudes, length, time, and mass" have been obtained, he says, we "can hence obtain a whole series of further equations for force, temperature, etc. by merely considering the dimensions of the quantity in question." He then tries to show how various physical relations, such as the ideal gas law, can be deduced from simple physical assumptions and his proposed principle of similitude. For relations about gravitation, however, a contradiction arises, which he embraces and uses to propose new criteria for an acceptable theory of gravitation. He ends feeling triumphant about his proposed principle. It's a new relativity principle, he concludes: the "principle of the relativity of size"!

> ... in the transformation equations which we have developed we have shown just what changes have to be made in lengths, masses, time intervals, energy quantities, etc. in order to construct such a miniature world. If, now, throughout the universe a simultaneous change in all physical magnitudes of just the nature required by these transformation equations should suddenly occur, it is evident that to any observer the universe would appear entirely unchanged. The length of any physical object would still appear to him as before, since his meter sticks would all be changed in the same ratio as the dimensions of the object, and similar considerations would apply to intervals of time, etc. From this point of view we see that it is meaningless to speak of the absolute length of an object, all we can talk about are the relative lengths of objects, the relative duration of intervals of time, etc., etc. The principle of similitude is thus identical with the principle of the relativity of size.

Einstein had shown that the principle of relativity of uniform motion led to the conclusion that it was meaningless to speak of two events occurring simultaneously—that one could talk only about relative simultaneity, never absolute simultaneity. Tolman structures his claim along the same lines, in saying that the principle of the relativity of size leads to the conclusion that it is meaningless to speak of absolute length. This seems at least on the face of it at odds with D'Arcy Thompson's view about the importance of size—that different laws govern at different size scales. We shall see that Tolman did not have the last word on the issue of the principle of similarity he proposed. Still, we may ask why he addressed the topic of the significance of size in physics at all. It is clear why a physicist would be interested in principles of relativity of motion, for relative velocities were essential to analyzing any dynamic situation in physics. But why an interest in size and scale?

It does seem that the question of the effect of size in scientific works was on people's minds even earlier, and when answered in one context, seems to rise in another. The question of size had been raised by Newton, who endorsed the notion of replica miniature worlds, at least in terms of their dynamic similarity, and by Galileo, who pointed out that, although there are certainly rules that inform us how to build replicas of different sizes, sometimes there are so many things you have to take into account in making a replica of a different size that you might not be able to address all the considerations: the rules are more complicated than just keeping relative magnitudes of the same kind the same. This is the point D'Arcy Thompson picks up on, and so there is at least an apparent tension between, on the one hand, Thompson's insight about how different life is for organisms that live in worlds at different size scales, and, on the other hand, his conviction that physical laws are universal, and that that universality has to extend across worlds of different size scales.

Actually, in Osborne Reynolds' most famous work, the 1883 "An Experimental Investigation of the Circumstances Which Determine Whether the Motion of Water Shall be Direct or Sinuous, and of the Law of Resistance in Parallel Channels," he expresses his conviction that physical laws are universal, when faced with an apparent challenge to it. In a subsection of the paper he called "Space and Velocity," he expresses this conviction and hints at an unexpected resolution to the seeming paradox:

> As there is no such thing as absolute space or absolute time recognised in mechanical philosophy, to suppose that the character of motion of fluids in any way depended on absolute size or absolute velocity, would be to suppose such motion without the pale of the laws of motion. If then fluids in their motions are subject to these laws, what appears to be the dependance of the character of the motion on the absolute size of the tube and on the absolute velocity of the immersed body, must in reality be a dependance on the size of the tube *as compared with* the size of some other object, and on the velocity of the body *as compared with* some other velocity. What is

the standard object and what the standard velocity which come into comparison with the size of the tube and the velocity of an immersed body, are questions to which the answers were not obvious. (emphasis added)

Reynolds considered this the philosophical, as opposed to the practical, aspect of his "experimental investigation" and the primary result of it. An idea about how the apparent paradox might be resolved had come to him only after he had figured out that the law of transpiration depends on an unusual relation of magnitude: "the size of the channel [the opening the gas has to flow through] and the *mean range* of the gaseous molecules." This discovery arose when he realized that a change in temperature affected the rate of transpiration, or gas flow, through the pores. Then he looked at Stokes' equation for further clues, upon which he realized that the form of Stokes' equation did contain the information that there was a relation of a sort that had been hitherto overlooked: a relation between what he called "dimensional properties"—properties that did depend on size (velocity, size of the tube) and "the external circumstances of motion." Deriving and then equating different expressions for fluid acceleration yielded what later became known as the Reynolds Number. He reported his result at the outset of the paper as follows:

> In their philosophical aspect these results relate to the fundamental principles of fluid motion; inasmuch as they afford for the case of pipes a definite verification of two principles, which are—that the general character of the motion of fluids in contact with solid surfaces depends on the relation between a physical constant of the fluid and the product of the linear dimensions of the space occupied by the fluid and the velocity.

This seems to be saying that the way to resolve the paradox between the conviction that size can only be relative, and our experience with phenomena that do seem dependent on size, is to expand the notion of "relative size" to include relations of linear magnitudes to other magnitudes. The crucial ratio is still dimensionless, as a ratio of two linear magnitudes would be, but it involves relations such as the viscosity of a fluid, not just the geometry of the situation. This still doesn't answer the question of whether there can be a miniature universe or not, for the question of what is meant by a miniature universe becomes more complex. Preserving the things that a length is relative to is no longer a matter of size, no longer merely a matter of geometry. Thus, it is natural that the question of whether a miniature universe can be a replica universe was bound to be asked again—as Tolman did, in 1914.

There may be some historical reasons for the intensified interest in scale, similarity, and transformation equations (correspondence rules) among physicists in the years leading up to 1914. Certainly there are different scales of magnitudes described by the kinetic theory of gases. Onnes's 1913 Nobel lecture in which he credits use of the law of corresponding states for his success in liquefying helium may have brought attention to the power of using scaling laws and principles of mechanical similarity, since they were involved in Onnes's derivation of that law. That alone would explain Tolman's interest: although he worked in a number of areas, he had a special interest in foundations of statistical mechanics, and he later authored a textbook on it.

An impetus for the interest in scale and similarity may have come from other areas of physics as well. For instance, did the discovery in 1911 that there was something much, much smaller than an atom—its nucleus—raise questions about what things were like inside the nucleus, or about what sorts of laws governed at such tiny scales? That is, did it raise the sort of image for physics that D'Arcy Thompson had evoked for biology: the image of how different things are in the world within a world within "our" world experienced by a bacillus? Or, alternatively, might the discovery of atomic nuclei have spurred reflection on, and helped renew commitment to, the universality of physical laws at all scales, as in Tolman's paper? Einstein's special theory of relativity emphasized the difference between cases in which velocities were very small in comparison to the velocity of light and those that were not. It was interesting that Einstein presented that theory as continuous with, rather than an overthrowing of, classical mechanics, and, emphatically, as a consequence of universal laws. The time around 1914 was also a time when there were new techniques allowing measurement of distances between molecules, meaning that assumptions that previously could only be debated as theoretical or inferred could be more directly determined experimentally.

Whatever the historical reasons for the interest, principles of similarity were in fact relevant to discussions and active research programs in fields as diverse as zoology and subatomic physics, and to endeavors as practical as applied aerodynamics. Another historical fact in 1914 was that the U.S. lagged behind almost every European country in aerodynamic research. Some in the U.S. were trying to change that. The ensuing bureaucratic struggle set the stage for the writing of Edgar Buckingham's remarkable papers on physically similar systems that had something to say in response to all these claims about similarity that were swirling around in 1914.

Chapter 7

Models of Wings and Models of the World

The paper by Stanton and Pannell on research into "similarity of motion" performed at Britain's National Physical Laboratory in the years just prior to 1914, which, they said, was research of practical value to "the prediction of the resistance of aircraft from experiments on models," was read to the Royal Society of London in January of 1914. Finally there was a comprehensive research study systematizing experimental knowledge about, and describing methods for making predictions from, fluid flow experiments performed over a wide variety of conditions. The 1914 paper provided systematized data from experiments on models in wind tunnels and validated the ranges over which they were valid. However, in the last chapter, we saw that we can tell from Rayleigh's later remarks in 1915, in which he was concerned to address skepticism about the validity of appealing to the principle of dynamic similarity in making predictions from model experiments in wind channels, that conclusions of the sort Stanton and Pannell drew in their "similarity of motion" paper were still met with less than full acceptance.

At least in Britain, the mathematical discipline of hydrodynamics and the practical discipline of hydraulics were still two separate disciplines in 1914. The first was exemplified by Horace Lamb's *Hydrodynamics*, a compendium of advanced mathematical solutions to differential equations; the second was exemplified by texts and handbooks such as A. H. Gibson's *Hydraulics and Its Applications*. Both Lamb and Gibson were professors at the University of Manchester for parts of their careers. Osborne Reynolds had retired from his post as an engineering professor in 1904, so he was no longer active there by the time Wittgenstein arrived in 1908, but he had been the founding member of the engineering school, and his influence permeated their activities. Gibson explicitly pays tribute to Reynolds as the major and overarching influence on his 1908 hydraulics text. The hydrodynamics-hydraulics divide was not a matter of differing local engineering styles; it was one of distinct disciplines that coexisted side by side in the same institutions. Reynolds managed to straddle both fields. He had spent a year in an engineering workshop in addition to his university degree from Cambridge. It was not at all uncommon for a scientist or mathematician to teach a course such as hydrodynamics, but Reynolds held a unique place in that he began his career at Manchester as a professor of engineering, rather than as a professor of physics or mathematics. In his case, something (perhaps it was the fact that he was seen primarily as an engineering professor, or perhaps it was because similarity was associated with hydraulics) seems to have hindered appreciation of the work he did in hydrodynamics as pure science. As Rayleigh noted, even after Reynolds' impeccable scientific work on hydrodynamical similarity, it was not accepted as readily as it should have been. There are vestiges even today of the divide in the world in which he worked; on the section of the University of Manchester's web site devoted to Reynolds, there are two separate sections, as if of a multiple personality: "Reynolds the Scientist" and "Reynolds the Engineer." The basis for the split is in part a matter of the different audiences for whom the different works by Reynolds were written. Reynolds belonged to both the Manchester Association of Employers, Foremen, and Draughtsmen as well as the Manchester Scientific and Mechanical Society, and he addressed different papers to the two different groups.

Reynolds had done mathematics at Cambridge. He had even been a wrangler for the mathematical tripos, that demanding multiday exam for which many undergraduates spent long days preparing with coaches. To place in the ranking as a wrangler was a distinction commanding respect. Thus, Reynolds showed he could hold his own and more on scientific topics. The mark of his work, though, was nothing so mundane as excellence in an established field; it lay in the extraordinary genius he displayed when it came to conceptualizing problems.

An example of Reynolds' ability to conceptualize practical problems so that their solution turned on fundamental principles was the work described in the preceding chapter, in which his thoughts took a quirky and brilliant path in a chain of reasoning that connected two seemingly unrelated phenomena: the transpiration of gases through porous walls, and the sudden formation of eddies in fluid flow in a pipe. Stanton and Pannell's 1914 paper, or any paper accomplishing the same thing for wind tunnel research, owed much to Reynolds' genius. To think Reynolds' discovery a matter of engineering skill or a straightforward consequence of a largess of knowledge about a particular subject matter is an egregious underestimation of the significance of what Reynolds called the "philosophical aspects" of the investigation that led to what is now known as the "Reynolds Number" (the dimensionless quantity whose value characterizes flow regime). He was right to call them philosophical, though the point is often missed: the intellectual path that led to that investigation from his earlier investigation into the transpiration of gases lay in detecting a paradox. He saw that being able to explain, not merely describe, the fact that transpiration could be induced by temperature difference as well as by pressure difference brought him to fundamentals about the laws of physics.

Reynolds saw that there was a paradox to be resolved in explaining the observed dependencies of the phenomenon of transpiration through pores in a plate using the kinetic theory of gases, without giving up his conviction that there cannot be anything such as absolute velocity. It's a philosophical, a priori, logical point. It is not something that belongs to one or the other of hydrodynamics or hydraulics. Boiling down all his varied observations, Reynolds concluded first that corresponding gas densities resulted in corresponding velocities through a given plate, and then, more generally, that "As long as the density of the gas is inversely proportional to the coarseness of the plate, the transpiration results correspond" and "In fact, the same correspondence appears with all the phenomena investigated." As we saw, the intellectual breakthrough in his analysis involved recognizing that a ratio involving length might involve another quantity such as velocity, whose dimensions are length divided by time. He found that the correct explanation of transpiration of gases involved such a nonsimple ratio, which allowed him to escape the paradox that arose when characterizing the gas properties in terms of molecular velocities and attempting to explain the phenomena in terms of the state of the gas alone. These insights about velocity—that there cannot be such a thing as absolute velocity, and that dimensionless ratios need not be simple ratios of like quantities—guided him several years later to the successful solution to the problem of the onset of turbulent flow at critical velocities.

Reynolds' predilection for using fundamental principles of science to achieve practical results and solve problems in hydraulics made him a non-scientist in the eyes of many. In contrast, his contemporary Rayleigh was unqualifiedly considered a scientist. In his 1915 paper, Rayleigh seems consciously aware of the respectability and weight his approval could bestow upon the use of wind tunnel model experiments. Rayleigh recognized not only the practical significance of Reynolds' work, but the sound scientific thinking behind it, and he sought to persuade others of the soundness of its basis, partly by arguing for it, but also partly by the weight of his authority in endorsing it.

There was distrust and reserve on both sides of this hydrodynamics-hydraulics divide. A tinge of defensiveness can be detected in A. H. Gibson's explanation of the subject matter of hydraulics in the 1908 preface to his text:

> Were water a perfectly non-viscous, inelastic fluid, whose particles, when in motion, always followed sensibly parallel paths, Hydraulics would be one of the most exact of sciences.

> But water satisfies none of these conditions, and the result is that in the majority of cases brought before the engineer, motions and forces of such complexity are introduced as baffle all attempts at a rigorous solution.

> This being so, the best that can be done is to discuss each phenomenon on the assumption that the fluid in motion is perfect, and to modify the results so obtained until they fit the results of experiment, by the introduction of some empirical constant which shall involve the effect of every disregarded factor.

As for the validity of using theoretical consequences of the obviously false assumption that water is an ideal fluid, Gibson both admits the dangers of doing so incautiously and justifies the appropriate use that can be made of such theoretical consequences. First, he warns the hydrodynamicist: "apart from these experimentally-deduced constants, his theoretical results are, at the best, only approximations to the truth, and may, if care be not taken in their interpretation, be actually misleading." But, to those who would criticize the theoretical treatment, he points out that "the results so obtained provide the only rational framework on which to erect the more complete structure of hydraulics."

It is easy to understand how the use of experimental engineering models got lumped in with everything else on the side of "Hydraulics." Although Stanton and Pannell's "similar motions" paper actually begins with references to Helmholtz's and Stokes' work using equations for non-ideal fluid flow, and although they make it clear that their work involved investigating "the conditions under which similar motions can be produced under practical conditions," the way that Stanton and Pannell present the experimental data collected from the National Physical Laboratory wind tunnel experiments could also be seen in terms of providing empirical constants that, as Gibson put it in his hydraulics text, "involve the effect of every disregarded factor." Stanton and Pannell presented a relation for "the resistance of bodies immersed in fluids moving relatively to them." One part of the expression, as we saw earlier, was an unspecified function of the product "v d/ v" (that is, (velocity × length) / (kinematic viscosity)) that was left to experimentation to determine. Thus, it is easy to think of the basis for wind tunnel experiments along the lines of other approaches in hydraulics, where one is essentially obtaining correction factors to solutions for ideal fluid flow. Of course, this is not quite what one is doing when using a method based on the principle of dynamical similarity, but the first uses of wind tunnels were not based on Rayleigh's principle of dynamical similarity. There really was no standard basis or methodology. There were well-developed and time-tested rules of correspondence for model experiments of naval ships, and so, by analogy, model experiments for airfoils were expected to be informative as well, though with correspondence rules of their own. These methods were generally valid only over limited ranges. It was known that using geometrically similar wing shapes and duplicating expected fluid conditions did yield useful data for designing and building aircraft, but the range of validity for applying the results was often unknown. In very early work with wind tunnels, experimentally determined correction factors of various sorts needed to be applied, and there may well have been an inkling of truth in the suspicion at the time that making predictions from model experiments was an applied art whose reliability lay more in the skill and knowledge of the practitioner than in the soundness of a method available to all.

The involvement of correction factors in Lilienthal's tables of aeronautical data that Chanute had passed on to the Wrights was so obscure and difficult to sort out that the Wrights thought Lilienthal's data was "seriously in error." They got around relying on Smeaton's coefficient by their ingenious use of a balance inside the wind tunnel. Their method was based on direct comparison of cases rather than raw data to be used in a calculation in which correction factors were compounded upon one another. As one historian put it: "[the Wrights] did mechanically what Lilienthal had done mathematically in the presentation of his data." In the end, as we have seen, their data matched Lilienthal's, but the fact that this was almost impossible for even them to see at the time shows how deeply embedded the data was in special cases and layers of experimentally determined correction factors. Stanton and Pannell used Reynolds' approach, based on characterizing and tabulating their experiments in terms of the nonsimple dimensionless parameter that Reynolds had used in characterizing fluid flow regimes. They extended this work so that it applied more generally, validating the approach that experiments in wind tunnels could be used to predict larger-scale situations since keeping the dimensionless Reynolds Number the same between model and large-scale machine meant that fluid motions, and hence aeronautical performance, would be similar between the model and the large-scale machine.

There had been a wind tunnel at the National Physical Laboratory since 1903, due to Stanton's efforts, and in 1909 the British Royal Aircraft Factory was established, in affiliation with the National Physical Laboratory. 1909 was historically significant to military aeronautical research, as it was the year that Bleriot flew across the English Channel separating Britain from continental Europe. Certainly many perceived this as a sign of a new vulnerability due to the advent of practical controlled flight, and hence of widespread support for a military fleet of flying machines. 1909 was also significant for another step Britain took: that year, Britain's Advisory Committee on Aeronautics was formed. This turns out to have consequences for our story, for it got together people ordinarily separated by discipline or profession, enjoining them to work on common problems: "Distinguished scientists and engineers from public and private life were appointed to the committee, which included representatives of the armed services, the Meteorological Office, and the National Physical

Laboratory, the government agencies most directly concerned with aeronautics." The role of the committee was "the superintendence of the investigations at the National Physical Laboratory and . . . general advice on the scientific problems arising in connection with the work of the Admiralty and War Office in aerial construction and navigation."

In taking these actions in 1909, Britain was just catching up with the rest of Europe. In 1908, Germany expanded its efforts in aeronautical research, treating it as a serious scientific venture: "The aero-dynamical laboratory of the University of Gottingen, established in the year of Wilbur Wright's first European flights, was . . . funded from external sources, including government, industry, and private associations, and was directed by Professor Ludwig Prandtl, with the advice of prominent scientists and engineers." Prandtl's work in interleaving experimental results such as measurements of the thickness of the layer of laminar flow that formed against a body in a fluid stream, and the flow conditions at which shedding of vortices in the surrounding stream occurred, were not just a matter of determining correction factors. He used experimental results to provide the boundary conditions that served as inputs to analysis that yielded mathematical solutions. Actually, Prandtl's appointment as head of a new research organization in 1908 was an outgrowth of a paper he had presented in 1904, not too long after details about the Wright Brothers' flying machine (including photos) had been leaked by an enthusiastic and well-meaning Chanute to aero clubs in Europe.

Prandtl was an ex-engineer-turned-professor in the Polytechnic at Hanover conducting research on air flow when he presented his paper at the Third International Congress of Mathematicians in 1904. It didn't make much of a splash—except with Felix Klein, then a prominent mathematician at the University of Gottingen. It was in the 1904 paper that Prandtl had laid out his plan of treating flow around bodies: What was distinctive was that he analyzed the problem into several distinct questions: what happened at the boundary between the "skin" that formed against the body, and what happened on either side of it—within the "skin" and outside it, within the main fluid stream. He showed that, in the mainstream, the beautiful mathematical solutions of vortex formation that were obtained by neglecting viscosity could be applied to even these real fluids. In the other part of the flow, under the "skin" formed around the body, viscosity did have to be taken into account. And, crucially, what happened in one part—formation of vortices in the mainstream—set conditions for what happened in the other, via setting boundary conditions at the interface between the two layers. Klein saw the potential of Prandtl's approach and brought him to a post in Gottingen right away. There, he was astounding.

According to a colleague: "his ability to establish systems of simplified equations that expressed the essential physical relations and dropped the non-essentials was unique . . . even compared with his great predecessors in the field of mechanics." Prandtl actually used a water tank for some of his most famous experiments. Instead of towing objects in the tank, though, he used a water wheel to move the fluid, much like fans were used to push air through wind tunnels. His results for airfoils were thus based on hydrodynamical similarity at a very basic level, for it is hydrodynamical similarity that justifies saying that, so long as the correspondences between relevant quantities could be established by dimensionless parameters, experiments with water could be used to predict the behavior of air. Prandtl's approach went beyond hydrodynamical similarity, though, for he combined mathematical solutions and experimental results in an unprecedented kind of synthesis. In building the new research organization at Gottingen, Klein took advantage of the treasure Germany had in Prandtl, in conjunction with the fruits from its educational institutions that had for decades produced highly educated mathematicians, scientists, and applied researchers.

Actually, England had talented visionaries as well, but less of an infrastructure capable of recognizing and using them. A case in point to compare with Prandtl is Frederick William Lanchester, who had been working on the theory of flight since before the turn of the century. In 1894, he gave a talk in Birmingham, England, in which he discussed his theory of lift and the soaring flight of birds, a review of Langley's experiments in flight, and power requirements for flight. Then, in 1897, he says, he submitted a paper on his own ideas about flight to the Physical Society of London, which was rejected. Ten years later, in 1907, he developed the ideas further and published a work titled *Aerodynamics* and, soon thereafter, one on stability in flight (*Aerodonetics*). By all accounts, there is some overlap between his work and Prandtl's. Lanchester did visit Gottingen in 1908, to discuss a German translation of his *Aerodynamics*. In describing the relation many years later, Prandtl said that, although the necessary ideas of the lift theory had occurred to him before he read Lanchester's *Aerodynamics*, "we in Germany were better able to understand Lanchester's book when it appeared [in 1907] than you in England. English scientific men, indeed have been reproached for the fact that they paid no attention to the theories expounded by their own countrymen, whereas the Germans studied them closely and derived considerable benefit there from." Lanchester's biographer, who remarks that "Lanchester's position is now secure as the originator of much of the vortex theory of wing lift and induced drag," still grants that "as a mathematical modeller, Lanchester does [not] rank in the Prandtl class. Doubtless this can be attributed, at least in part, to deficiencies in the training of British engineers, deficiencies not present in the Continental system in which sophisticated mathematical modelling and advanced techniques of mathematical analysis had long been taught."

So the response in Germany, especially the establishment of the lab at Gottingen in 1909, was a definite and vigorous step toward treating aeronautics as a subject to which pure mathematics and physics could really be applied. In terms of technology, too, Germany led: the wind tunnel built at Gottingen in 1908 under Prandtl's direction was unlike any other. It had many advanced features: "vanes at the corners to turn the flow" and "strategically positioned screens and honeycombs to homogenize and quiet airflow," and it was "the world's first continuous-circuit, return-flow wind tunnel." In 1913, Richard von Mises, educated at the Technische Hochschule in Vienna, Austria, and professor of applied mathematics at Strasburg gave a university course on "the mechanics of flight." These lecture notes were soon developed into a small textbook, *Fluglehre*, originally published in German, and eventually published as a large compendium that was translated into English in 1945 and was still in print decades later. Thus, by 1914, Germany and the German-speaking world were firmly on a trajectory that would give them the lead in aerodynamics for some time.

The field of hydraulics was becoming less polarized in Germany: In 1909 a smaller and less well-known hydraulics lab was developed by Alexander Koch at the Polytechnic Institute at Darmstadt; Koch's view expresses the sentiment growing in Germany at the time: "We need . . . in place of mathematical hydrodynamics and empirical hydraulics a simple, clear and practical hydrodynamics." In 1914, the first edition of the Vienna-born German professor Philip Forchheimer's *Hydraulik* appeared. Rouse and Ince, writing in 1957, said that it was then still "the outstanding compilation of and commentary upon hydraulic data of all time."

Russia had established a laboratory just after the Wright Brothers' successful flight at Kitty Hawk in 1903: a "privately owned university-connected laboratory was established in Russia in 1904 when the Aerodynamic Institute of Koutchino was appended to the University of Moscow." The private funding came from a scientist, D. Riabouchinsky; although the Western world would not know it until later, Riabouchinsky also proved a mathematical theorem about the method of similarity in 1911. The Koutchino wind tunnel in Russia produced what has been judged "excellent technical data" for many years. It was in use in 1914 and would be for years afterwards.

The French, however, with their longstanding love of aeronauts and balloons, had always thought of themselves as the country of machines for human air travel. They were the ones who had sponsored frequent exhibitions, competitions, and aero clubs, making France a geographic magnet for aeronauts and inventors of flying machines, granting them celebrity. It is fair to say that they had first led the way in aeronautical research, too; by NASA's official account:

> Most of Europe's early aeronautical laboratories were in France . . . Work at the Central Establishment for Military Aeronautics at Chalais-Meudon near Paris was complemented by the researches of Gustave Eiffel, working at his famous tower between 1902 and 1906, then in laboratories at the Champ de Mars and in Auteuil . . .

After graciously granting that the 1908–1909 European exhibitions showed beyond all doubt that the Americans really had won the technological race for the solution to the problem of controlled flight, the French continued and expanded their aeronautical research:

> . . . after 1912 at the privately endowed Aerotechnical Institute of the University of Paris located at St. Cyr. Like similar organizations to follow, the Institute had a director supported by an advisory committee composed of scientific and aeronautical experts from the University of Paris, the Aero Club of France, and government departments concerned with aviation.

Meanwhile, across the ocean in the United States of America, the very country in which the problem of controlled, powered heavier-than-air human flight had been solved by the Wright Brothers, the U.S. government was going in the opposite direction. Some wind tunnel and aeronautical research facilities had actually been shut down, and, though the Smithsonian had just reopened the Langley Aerodynamical Laboratory, proposals for funding from Congress and efforts to form a U.S. Advisory Committee on Aeronautics seemed to be going nowhere. It was only in 1913 that Zahm's wind tunnel, located in the Washington Navy Yard, went into operation. The reason these efforts, some begun in earnest as early as 1911, were still going nowhere as late as 1914 is a long, complicated story of bureaucratic competitiveness (between the Navy, the Smithsonian, and the National Bureau of Standards) and legalistic minutiae, but a U.S. Advisory Committee on Aeronautics modeled closely on Britain's was finally established early the next year through a last-minute legislative trick. The result of that little trick (tacking authorization for its formation as a rider onto a Navy appropriations bill), though it provided hardly any funding, became, in time, the very large, very sophisticated, heavily funded organization known today as NASA. When a history of NASA was written in 1976, it skipped over the year 1914, moving from the frustration and anxiety of 1913 to the meager opportunity to do something about it finally gained in 1915:

> In 1913 the clouds of war were gathering over Europe—and casting their shadows on America. The European powers were reacting to arm themselves against each other—not only with conventional land and sea armaments, but also with the new weapon of the twentieth century—the airplane. In their race they overcame the U.S. lead established by the Wright brothers and left [the U.S.] in a technological backwash. Particularly disturbing to American observers was [the U.S.'s] primitive and unorganized aeronautical establishment—a frail shadow of the research facilities and government subsidized industries arising in Europe.

> Most active among the small group of concerned men in the United States was the Secretary of the Smithsonian Institution, Charles D. Walcott. Convinced that the situation called for Federal sponsorship of an aviation organization, he worked hard selling the idea both

inside and outside the Government. After several false starts, he succeeded. On 3 March 1915, President Woodrow Wilson signed into
law a Navy appropriations bill with a rider establishing an independent Advisory Committee for Aeronautics.

The details of the Smithsonian's Walcott's having to work hard at "sell-
ing the idea," of the "several false starts" he endured, and of the efforts of
those who preceded him, are laid out in a later account of NASA's early history. On the one hand, the account seems a story of a tedious fight for turf
control of government-sponsored aeronautical research between the Navy
and the Smithsonian Institute, and so hardly relevant to our story. But
underlying the sentiments that made for strong allegiances to one side or
the other is something that *is* relevant to our story: attitudes toward the
relative importance of science and practical skill in aeronautical research.
And these were embodied in attitudes and beliefs toward the two most
famous American research programs in early heavier-than-air flight so far:
Samuel Langley's aerodrome and the Wright Brothers' flyer.

There was a concerted effort in 1914 to change public perception
about the Wright Brothers' claims of being the first to fly and about
Samuel Langley's failure to fly at all—and, with it, attitudes toward the role
science had had in inventing flight. Actually, the Wright Brothers were very
scientifically sophisticated, well-informed aeronautical researchers who
also happened to be very skilled at making sophisticated machinery with
the high-precision machine shop equipment they owned and used to make
custom high-performance bicycles of their own design. However, the perception was that they were "mechanics" and that the contrast between
Langley and the Wrights was a contrast between science and practical
know-how.

Langley and the Smithsonian were identified in the public mind, in
spirit as well as in formal affiliation. In James Tobin's chronicle of early
flight, in which he lays out the story of Samuel Langley and the Wrights as
a race of sorts, Tobin describes the Smithsonian in 1903 as "the best-
endowed, most prestigious institution of science, culture, and learning in
the entire nation" and Langley as "the most prominent scientist in the
United States" in public stature and prestige. The Smithsonian's Tom
Crouch writes that, when Langley was recruited by the Smithsonian in
1887, he:

had become a leader of American science. He held honorary degrees from some of the great universities of the world, including Oxford, Cambridge, Harvard, Yale, Princeton, Michigan, and Wisconsin. International scientific societies had showered him with awards. In 1886 he received the Henry Draper medal from his fellows of the National Academy of Sciences. The Janssen medal of the Institute of France, the Medal of the French Astronomical Society, and the Rumford medals of both the Royal Society of Great Britain and the American Academy of Arts and Sciences attested to his distinguished reputation.

Prestige in the scientific community was important to the Smithsonian. Tobin tells of reports that, when things had gone most terribly wrong in a public way with Langley's aerodrome in 1903, and he began losing respect in the scientific community, "a number of scientists undertook a quiet campaign to remove Langley from his post as secretary on the ground that his mind had given way, that he was endangering the fair name of the Institution by a series of foolhardy experiments which could never result in anything" although "[b]ehind the scenes, . . . [friends] fought on his behalf and quashed the attempted coup."

After the Wrights' very public and astoundingly successful exhibitions in 1908, when the practicality of manned flight was no longer in question and Langley's belief in the possibility of manned flight thus was vindicated, the Smithsonian contrived a way to share in the glory by giving credit to Langley's vision:

[Alexander Graham] Bell helped to inaugurate a program by which the Smithsonian would award an annual Langley Medal for "specially meritorious investigations in connection with the science of aerodromics and its application to aviation." The Wrights were chosen as the first recipients, and two identical gold medals were struck at the Paris Mint. Yet when Bell was chosen to speak at the awards ceremony at the Smithsonian on February 10, 1910, the inventor barely mentioned the recipients' names. He congratulated the Wrights for bringing "the aerodrome" [Langley's word for a manned flying machine] to "the commercial and practical stage," just as Langley had predicted someone would. But it was Langley himself, Bell declared, who must be recognized as "the great pioneer of aerial flight"; who had divined

"Langley's Law" which "opens up enormous possibilities for the aero-drome of the future"; and who had constructed "a perfectly good fly-ing machine" that failed to fly only because of its faulty launcher. "Who can say what a third trial might have demonstrated?"

This view, already well rooted among Langley's friends, spread to others in aeronautical circles, especially those who resented the Wrights' insistence on claiming a financial reward for their work.

Or, perhaps the view was hatched out of resentment to begin with, a resentment of what seemed to Langley's friends the cruelty and unfairness of fate. Langley had been working on flight long before the Wright Brothers took it up, he had approached the study systematically, and he had had success with powered aerodrome models—just not ones big enough to carry a human. Then, it seemed to Langley's friends, finally, just when it was all about to pay off in 1903 with his full-scale human-carry-ing-sized version, a tiny hardware glitch had kept his flying machine from taking off, and Langley's funding was cut off as a result. Meanwhile, as they saw it, the Wright Brothers had come along and, benefiting from all Langley's scientific research, had tweaked their way to success and glory. As one history of NASA tells it, that the Wrights' achievement "flowed as much from broad study and scientific method as from their natural intu-ition and genius" was not realized at the time, and so

> the Wrights were viewed by many as mere bicycle mechanics, and their achievement as a fortuitous victory over their nearest American rival, Samuel Pierpoint Langley, secretary of the Smithsonian Institution and scientist of flight. By extension, it was a victory over science itself. Or so it seemed.

By 1914, this resentment had turned into outright manipulation to make the facts fit these attributions of priority to Langley. To shore up the view that Langley had in fact constructed "a perfectly good flying machine" and hence that science had not been bested by engineering, Langley supporters decided to give Langley's aerodrome the chance that, they felt, fate had so cruelly withheld. Alfred P. Zahm, who was also a paid expert witness to Glenn Curtiss in a lawsuit with the Wrights over patent infringement, had by then come to resent the Wrights for reasons of his own. Zahm had been a physicist and professor at Catholic University near

Washington, D.C., and one of the first American scientists to perform research in wind tunnels; he had founded and run a very large wind tunnel with advanced features and very sensitive instrumentation. It was financed by a wealthy industrialist who believed in the future of flight; when the industrialist died in 1908, the facility closed. Zahm had done very good original work with that advanced wind tunnel and had made new discoveries about airflow around dirigible air hulls. Of course, he thought funding research at a U.S. aeronautical laboratory with a wind tunnel was urgently needed. He had just designed the new one at the Navy Yard. So Zahm had good reason to worry about public perceptions that might hinder support for laboratory research, such as the misperception that the Wright Brothers' success meant that engineering had bested science in flight research. From how things looked to him then, vindicating the worth of science in flight research meant vindicating Samuel Langley.

It would suit both Zahm's interest in seeing Langley's reputation restored and Curtiss's interest in undermining the Wrights' claim to priority if Langley's 1903 aerodrome could be shown capable of sustained, controlled, manned flight. Hence, the 1903 aerodrome, which was stored at the Smithsonian, was brought out for another trial. Zahm and Curtiss gave the 1903 aerodrome another chance. They performed a test at Lake Keuka to show it could at least become airborne, and they took photographs of the brief moments after it had been lifted above the surface of the water as evidence. But that is not all they did. They then modified it—a lot. Among other things, Curtiss "refitted the wings to improve their lift-to-drag ratio. He also modified Langley's propellers for greater efficiency." The aerodrome, thus modified, was capable of sustained flight. It was the results of these flights with the modified aerodrome that were publicized, thereby enabling the Smithsonian to replace the image of the failed aerodrome flight of 1903 in the minds of the public with one of a 1914 success. It meant that those proposing that a scientific research institution take the lead on aeronautical research no longer need hang their heads in shame about the disappointing outcome of the large amounts of money that Congress had already spent on Langley's aerodrome.

Orville Wright had his suspicions about the 1914 aerodrome trial vindicating the Smithsonian's claims that Langley should be given credit for inventing the first practical flying machine, but he held his tongue. In 1914 he thought Walcott and the Smithsonian had been misled, but, according to one historian, it would not have been easy for him to tell them so: "He had admired Langley, who had died in 1906, and did not want to raise criticisms that could be seen as attacks upon the dead. He was not certain of the extent of Curtiss's modifications." As for fraud on Walcott's part, Orville had such a high regard for the Smithsonian that "he was unprepared to assert that its officials knew what Curtiss had done." Things would change years later when Orville found out the Smithsonian's Walcott was not so innocent as he had thought, but in 1914 this is how things stood. With the worth of Langley's aerodrome research now settled (it seemed to him), Walcott could deflect criticisms about funding scientists to do flight research that cited the aerodrome's 1903 failure. It turned out there were other obstacles to Congressional funding for an aerodynamics laboratory, and in the end what Walcott got Congressional approval for was a U.S. analogue of Britain's Advisory Committee on Aeronautics: a committee to advise Congress on aeronautical research.

That was the background against which the U.S.'s Advisory Committee on Aeronautics was proposed. For several years before it was actually formed in 1915, various organizations were proposed and rejected, against differing views of the role physicists ought to have in flight research. So there was actually a lot of interest in and attention to the scientific basis for methods of aeronautical research in the years just prior to 1914. The Smithsonian sent two representatives (Alfred P. Zahm and Jerome Hunsaker) to tour European research facilities in 1913, and their report was presented in 1914. Britain's organizational structure for aeronautical research, which had a government-sponsored research laboratory and an advisory committee consisting of people of a variety of professional, scientific, and bureaucratic associations directing and overseeing aeronautical research, was thought especially well-suited for the future American organization they hoped would be formed.

The physicist Edgar Buckingham was at the U.S. National Bureau of Standards during the struggle for future control between the Smithsonian, the National Bureau of Standards, and the Navy over the hoped-for government aeronautical laboratory. He would later serve on the Advisory Committee on Aeronautics. But he had a double life. As a physicist well-versed in foundational issues in thermodynamics, he had read Tolman's

1914 article in the *Physical Review* proposing the principle of similarity as an expression of a principle of the relativity of size. Although Tolman's article discussed a thought experiment in gravitational theory and mechanics, Buckingham could hardly have missed the bearing that the claim about observationally equivalent miniature universes had on model experiments.

Like Reynolds before him, Buckingham was put in a position where he had to straddle the scientist-engineer divide, and he made it his business to become knowledgeable in the field of aeronautical engineering. Zahm and Hunsaker's report of the European tour of aeronautical research facilities and organization that they undertook on the Smithsonian's behalf would have brought him up to speed on the work being done at Britain's National Physical Laboratory and alerted him to the work that Stanton and Pannell were then doing that was then reported in their 1914 paper on "similar motions." Buckingham had little previous experience or background in aeronautics until he began working on it around 1911 (which is when the first efforts to establish a government agency devoted to aeronautical research began in earnest). In a letter to Rayleigh in 1915, he explained the origins of his own 1914 paper on physically similar systems:

> Some three or four years ago, having occasion to occupy myself with practical hydro- and aerodynamics, I at once found that I needed to know more about the method [of dimensions] in order to use it with confidence for my own purposes. Since you and the few others who have made much use of the method of dimensions have generally referred to it somewhat casually as to a subject with which everyone was familiar, I supposed that the hiatus in my education would be easily filled. Accordingly, upon looking through your collected papers, the "Sound" [probably a reference to Rayleigh's "Theory of Sound"], Stokes's papers, and a few standard books such as Thompson and Tait [Principles of Mechanics] and Routh's Rigid Dynamics I was amazed at my failure to find any simple but comprehensive exposition of the method which could be used as a textbook. . . . Each one of your numerous applications of the method seemed perfectly clear, and yet their simplicity gave them the appearance of magic and made the general principle rather elusive.

It is telling that, in his search for an exposition of the theory of dimensions, Buckingham mentions looking at the main *mechanics* textbooks used in Britain, rather than in either Lamb's *Hydrodynamics* or Gibson's *Hydraulics and its Applications*. The books he did consult, Thompson and Tait's mechanics text and Routh's dynamics text, were the basic texts on the subject that a physicist would have on his shelf. But, then, Stanton and Pannell's report on the wind tunnel research at the National Physical Laboratory in January 1914 in their "similar motions" paper treats the subject from a physicist's point of view, too. They don't refer to texts used in engineering schools, such as Lamb's compendium on hydrodynamics or Gibson's on hydraulics. They use Rayleigh's equation for fluid resistance, they refer to Newton's *Principia* for the ratios and relations that are supposed to yield similar motions, and they refer to Stokes' and Helmholtz's work.

Approaching aerodynamics from the point of view of a physicist, and hence on the scientist side of the scientist-engineer divide, was not only consistent with the kind of community in which Buckingham worked and had been educated, but it was also the more politically acceptable image of aerodynamics to cultivate in 1914. That Buckingham was acutely aware that physicists and engineers were in different communities is evidenced by his remark to Rayleigh about his *Nature* article on the principle of dynamical similarity; he wrote Rayleigh that "a note, such as the one in *Nature* of March 18th, which has your authority behind it, has an effect far more important in the present state of affairs than any detailed exposition of the subject, however good, because physicists will be sure to read it."

Buckingham's "area of expertise," as they say, was thermodynamics. He didn't view thermodynamics in its most general sense as just a subspecialty in physics, though, but rather as an enlightened view of science in which thermodynamics encompassed all of classical mechanics. In his 1900 book *Outline of a Theory of Thermodynamics*, he wrote:

> Thermodynamics, . . . aims at the study of all the properties or qualities of material systems, and of all the forms of energy which they possess. It must, therefore, be held, in a general sense, to include pure dynamics, which is then to be looked upon as the thermodynamics of systems of which a number of non-mechanical properties are considered invariable. For 'thermodynamics,' in this larger sense, the more

appropriate name 'energetics' is often used, the word 'thermodynamics' being reserved to designate the treatment of problems which are
directly concerned with temperature and heat.

These remarks are interesting for several reasons. First, that
Buckingham sees "energetics" as a more generalized description of what
"pure dynamics" is indicates a tendency to seek out more generalized versions of theories with which he is already familiar. That is, he says that
"pure dynamics" is a special case where some of the things that would
appear as variables in the most general description of the system are
treated as constants within the subdiscipline of physics called dynamics.
This is further clarified later, wherein which variables are considered variable and which invariable is relative to how the "given system" is defined.
Thus, Buckingham's approach toward formalizing physics in his 1900
book on foundations of thermodynamics was to make the formalism he
proposed to use as flexible as possible and to build as few assumptions as
possible into the formalism. Conversely, in generalizing the existing science of dynamics, he chose to regard as variable certain variables that are
considered invariable in dynamics.

Such a choice is notable in that it reflects a tendency to question
whether constants could be other than they are, and it raises the question
of whether anything is really a "constant." This step of analysis is prior to
the application of the theory of dimensions. Applied to an analysis of
equations, these two tendencies would drive one to search for the most
general form of an equation. An analysis that allows one to regard a given
variable as constant or not, depending on how a system is regarded,
requires a correspondingly general treatment. This general treatment leads
one to rules that are less and less about a specific science and more and
more about the logic and structure common to multiple sciences. It calls
to mind Frege's remark that the laws of logic are the laws of the laws of
science.

Buckingham had studied in Leipzig, Germany for his doctorate in
1894 under Wilhelm Ostwald, who was also heavily involved in exchanges
about philosophy of science and foundational issues in science. Ostwald is
generally attributed with founding a new science, the field of physical
chemistry. Ostwald, as I recounted, was a professor at Leipzig and was a

personal friend and sometimes colleague of Boltzmann. His relation to Boltzmann, as far as the public was concerned, was to be on the opposing side of the debate over "energetics" and the use of models in physics. Ostwald was one of the most active proponents of energetics, a view on which all of physics could be treated in terms of energy principles, and he argued against the use of models. Here he meant models of unseen particles whose motions are supposed to explain the workings of heat and other properties of gases. Instead, he thought the equations of physics and the concept of energy were all that were needed, and that potentially misleading models could be dispensed with. The laws of thermodynamics, of course, are about balances of energy and work. In his *Principles of Mechanics*, Hertz had pointed out one advantage of energetics—the promise of a formulation of physics in which forces need not be postulated, since energy becomes the primary notion. Hertz proposed his own account, on which he took the insight that energetics offered, but supplemented it with his own account of matter. Like most other texts on Lagrangian mechanics, Hertz talks of systems that have corresponding values, such as corresponding velocities, and of the value of knowing the number of degrees of freedom of a system before choosing coordinates in which to characterize the system.

Even without bringing in principles of thermodynamics, energy approaches to mechanics and dynamics arise in the alternative formulation of mechanics known as Lagrangian mechanics. To those whose education in mechanics and dynamics begins with Newtonian mechanics, in which forces are identified and represented mathematically as vector quantities, with spatial directions and locations at a point at which they act on an object or particle, finding out about the alternative methods of Lagrangian mechanics is a liberating jolt. It is an unforgettable moment in an engineer's or physicist's education. For one thing, as mentioned earlier, Lagrangian mechanics allows formulating an equation in more generalized coordinates: what is determined is how many coordinates are required, which, in this generalized form of mechanics, are known as degrees of freedom. That is, what is determined is just how many coordinates will be needed to characterize the state of the system. Which coordinates or degrees of freedom they are is not determined, and some freedom of choice is allowed. Also, instead of the Newtonian laws and equations, which require detailed accounting of forces and accelerations along each of three spatial axes, in Lagrangian mechanics energy equations, rather than force balances, provide the means for predicting states and properties of a mechanical system. Besides the flexibility that the "new science" of Lagrangian mechanics allows in the conception and solution of problems, the existence of such very different alternative representations of the same situation, and the existence of two very different means of solving the same problems in dynamics, also stimulate reflection on questions about what is essential to the description of a situation in general.

These are just the kinds of questions Buckingham had been driven to reflect on his whole life. He brought a penchant for asking these kinds of foundational questions to the field of aeronautics, though as a mature scientist coming upon an emerging field for the first time. It seems that no matter how practical the subject matter he was dealing with, he approached it with the eye of a pure scientist, and one interested in foundational questions at that. Even among his colleagues at the Bureau of Standards, he was someone who forged his own way. It seems to have been his way, no matter what he turned his attention to, to dig deeper into foundations and then to rise to a more general view. When he served as a physicist at the Bureau of Soils (BOS) in the U.S. Department of Agriculture, working with Lyman Briggs:

> Buckingham [was] more completely focused on physics-based research. In Buckingham's 3-[years] on solid physics as the BOS, his achievements include one of the biggest single steps toward the physical quantification of soil-water flow (Buckingham 1907). Supported by his newly developed theory and experimental evidence, Buckingham introduced the concept of capillary potential . . . as an essential measure of the energy of soil water relevant to flow.

The problem bears some resemblance to the phenomenon of transpiration that Reynolds worked on. Explaining that problem involved identifying scaling effects as well as the proper correspondence rules for similar velocities and other transpiration phenomena. Reynolds had closed his paper on the transpiration of gases by remarking on processes in nature in which the actions that occur in the process are ones in which "the actions only become considerable within extremely small spaces," and he mentioned processes in "the interstices of soil." Twenty five years later, Buckingham realized the connection of the phenomenon of transpiration of gases to the questions about soils he was working on, and in his paper he refers to the phenomenon as "transpiration." Although he may not have known of this specific paper by Reynolds, his general conclusion is of the same sort as Reynolds': "the speed of diffusion of air . . . through these soils was . . . determined in the main by the porosity of soil." We know from Reynolds' comments that he found the work very intellectually stimulating and suggestive of new avenues of general significance, for it was this work that led him to discover that some properties of a gas depend on the gas density, and to think in terms of the kinds of nonsimple ratios that later led him to discover what is now known as the Reynolds Number. What Buckingham did on the question of aeration of soils had a similar general- ity, and he showed how experimental results could be used to infer results for any given case by thinking in terms of laws of correspondence. Briggs' letter submitting Buckingham's paper "Contributions to Our Knowledge of the Aeration of Soils" for publication said:

> This paper presents for the first time definite information regarding the rate at which a gas escapes by diffusion from the soil into the atmosphere, or vice versa. It is shown that the rate of diffusion varies approximately as the square of the porosity of the soil, and that the diffusion follows the laws for the free diffusion of gases. It then becomes possible to calculate the rate of aeration in any particular soil from results obtained in experiments on free diffusion. ("Letter of Submittal," Lyman J. Briggs, Physicist in Charge, November 8, 1904, front matter of "Contributions to Our Knowledge of the Aeration of Soils," Bureau of Soils, Bulletin No. 25, U.S. Department of Agriculture, Washington, D.C., Washington Government Printing Office, 1904.)

In spite of Buckingham's show of brilliance and the fact that the results he obtained provided a practical means of obtaining useful infor- mation, the scientist side of the scientist-engineer divide was not always appreciated in the U.S. Department of Agriculture. The kind of theoretical work that Buckingham had done in Briggs' lab was criticized by the Chief of the BOS in remarks directed toward Briggs. He was concerned that Briggs was "wasting" his time and ability in the BOS position, because "perhaps our problems are so difficult that they could not be treated in the strictly mathematical way in which your training has induced you to look upon them" and that "We have to depend on more crude methods of experimentation to formulate first approximate facts and laws before we can ever hope to apply rigorous mathematical physic (sic) measurements."

This general complaint, aimed at the use of pure science itself in the soil physics laboratory, arose in the context of the BOS Chief's complaining that too much time and effort was spent on outside work that was considered unrelated to the practical problems of the Bureau of Soils. One of the things Buckingham was doing in addition to developing a theory of the behavior of soil water from energy considerations for the Bureau of Soils was following debates in the scientific literature about the foundations of thermodynamics. In 1904, Buckingham and a relatively unknown physicist working in a government agency on the other side of the Atlantic Ocean— Albert Einstein—weighed in on discussions about the same article in the British journal *Philosophical Magazine*—an article about foundations of the laws of thermodynamics. Buckingham's discussion regarded the logical dependencies between certain thermodynamic laws. Einstein took note of Buckingham's commentary and wrote a summary in German of Buckingham's article for a German-language scientific journal. In the aftermath of the BOS Chief's transfer of Briggs (almost immediately after complaints about the sort of work Buckingham carried out), Buckingham left the BOS for the National Bureau of Standards. But he kept up his work on foundations of thermodynamics. He published in both American and British journals, often publishing an article first in an American journal and then a very similar one in a British journal.

One of his papers on foundational issues in science is especially noteworthy—"On the Time Scale," published in *Philosophical Magazine* in 1912. It indicates his continued and abiding interest in asking what is essential to a quantity in contrast to what is conventional (a matter of choice in how we choose to regard it). Buckingham argues that time scales are analogous to temperature scales, in that

> *in the nature of things* time is not a measurable quantity; . . . times can be numbered in accordance with any system we please which does not introduce ambiguity into the meanings of "before", "after", and "at the same time". . . . no system has any more justification than another, a priori.

The choice of time scale used in physics—"mean solar time"—is based on its compatibility with Newton's first laws. "In effect, we use the first law to fix our time scale." It's not a matter of saying anything about time, he says. "It would be an utter absurdity to say that we 'assume' that time 'really' does progress in this way." Rather, fixing the time scale is a matter of defining time: "We define our time scale by saying that Newton's laws shall be true, simply because it pleases us to do so; and it pleases us because we prefer simple to complicated statements of fact, when our choice is free."

This is not the same point that Einstein made in his 1905 paper on the special theory of relativity. There, the point about time was that there was no such thing as "really" simultaneous; Einstein had shown there that simultaneity was relative to an observer's situation. Here, Buckingham is talking about observers who are "situated as we are with relation to the two events," so the issue isn't one of absolute versus relative simultaneity. Rather, he is critical of "nonsense about clocks 'running at a constant rate,'" for, he says, the term "constant rate" is "quite meaningless except in connexion with some time scale outside of and independent of the clock itself." The bottom line is that our measure of time is a matter of which laws we want our notion of time to be compatible with:

> The notion of time, as defined by reference to Newton's first law and to the measureable quantity length, is so ingrained in us that the symbol "t" in our equations seems to us as if it ought, somehow or other, always to stand for the old familiar mean solar time. In reality it stands for a derived quantity which may be defined by means of true, i.e., measurable quantities, through the mediation of Newton's laws or Maxwell's equations or any other statement of observed facts which we may happen to be interested in having appear in a particular form. In the past we have paid attention only to Newton's laws, defined the time scale by reference to them, and used the same time scale in all branches of physics.

But, he argues, it is known that if we want to use equations (such as Maxwell's equations) for electromagnetic phenomena for cases in which the ratio of the velocities involved to the speed of light in a vacuum is appreciable, as it is in electromechanical phenomena, we should use those laws, rather than Newton's, to define the time scale for time as it appears in those equations. The question then remains as to whether "a quantity can be so defined that it may replace our familiar 't' in all the equations of physics, for observations made under all possible circumstances, and without requiring any too serious modifications in the form of the equations." Buckingham seems to be objecting to the tendency to ignore this question, for he closes the paper with the admonition that this problem cannot be settled "by any use of clocks which run 'uniformiter,' when 'uniformiter' is defined only by reference to a Latin dictionary."

This bird's-eye view of the relation between measurable quantities and scientific laws, we will see, distinguishes Buckingham's work on the foundations of model experiments from most other accounts of the method. It also differs from the view expressed in 1904 by Edward Nichols (editor-in-chief of *The Physical Review*) at the Congress of Arts and Science held at the 1904 World's Fair in St. Louis (as discussed in Chapter 3). Nichols, though born in England, was educated in America at Cornell and in Germany at Gottingen. He was asked to give a lecture in the "Physics" section, and he lectured on "The Fundamental Concepts of Physical Science." Recall that, in that lecture, he employed what he called "dimensional formula" in reasoning about concepts in a variety of the newer fields in physics. His view was very conservative: he considered the fundamental concepts of physics to be set and universal conceptions, of wide applicability, in terms of both subject matter and scale: "The science of physics . . . has for its foundations three fundamental conceptions: those of mass, distance, and time, in terms of which all physical quantities may be expressed." What he said about the inability to reduce the concept of "charge" to a combination of these three concepts is that "Not all physical quantities, in the present state of our knowledge, can be assigned a definite dimensional formula, and this indicates that not all of physics has as yet been reduced to a clearly established mechanical basis. The dimensional formula thus affords a valuable criterion of the extent and boundaries of our strictly definite knowledge of physics."

The physicist chairing the session in which Nichols' talk on dimensional formula was given was none other than Henry Crew, the physics professor whose translation of Galileo's *Two New Sciences* into English was published in 1914—a work, we have seen, in which the importance of size is argued for by using dimensional considerations. In general, when dimensional considerations or dimensional homogeneity were discussed, as in Rayleigh's use of "the method of dimensions," the quantities basic to physics were stated, but they were not critically examined as to whether they met criteria to be fundamental quantities. The same view is reflected in the discussions on similarity in Gibson's text on hydraulics in 1912 and in Stanton and Pannell's 1914 "similar motions" paper. It is explicit in D'Arcy Thompson's *On Growth and Form* and implicit in Galileo's discussions of physical similarity in *Two New Sciences*.

Not everyone in the field of physics took the extreme view that Nichols did—that all possible quantities could be expressed in terms of mass, distance, and time—but it is fair to say that even among those who allowed for more than these three fundamental quantities, homogeneity of dimensions was generally exhibited by decomposing each quantity into a set of fundamental ones (three if mass, distance, and time are considered sufficient; more if quantities such as temperature and charge are also considered fundamental quantities that can't be reduced to some combination of the other three). Whether temperature should be considered a fundamental quantity or whether (especially in light of the kinetic theory of gases) it was at least possibly reducible to some combination of the mass, distance, and time was a live question and was debated among physicists.

Buckingham wrote about physically similar systems and the theory of dimensions in a number of papers, the first two of which appeared in 1914. We have seen that there was also a blossoming of interest in similarity principles in a variety of disciplines and contexts in 1913 and 1914. It is likely that his interest was kindled in part, too, by the sense of urgency that existed at the time over establishing U.S. aeronautical research facilities, including a wind tunnel, for we know that he was reading reports from Britain's Advisory Committee on Aeronautics as early as 1911 or 1912. He may even have been assigned the task of working on standards for instrumentation in wind tunnels in his role at the National Bureau of Standards. Although that would be reason enough for his interest, there were, as we have seen, other reasons for a physicist to be interested in similarity for its own sake.

We know that Buckingham took note of the physicist R. C. Tolman's January 1914 paper in *The Physical Review* proposing a "principle of similitude" as a new principle of relativity—the relativity of size—for he included a response to Tolman at the end of one of his 1914 papers (the October 1914 *Physical Review* paper on physically similar systems). As a philosophically-minded physicist with one foot planted firmly on the scientist side of the scientist-engineer divide, and especially as a mature physicist with experience in seeing new physics in even the most practical problems (such as the behavior of mud and dirt), the thematic connection between the methodology of experimental scale models and Tolman's claim about similitude could hardly have escaped him. Also, the "law of corresponding states" in thermodynamics, which Onnes had just talked about in his December 1913 Nobel lecture, explaining that he had looked for its explanation in Newton's notion of mechanical similarity, would have brought up the topic of similar systems in the mind of anyone following the foundations of thermodynamics as well.

The first piece Buckingham published on the topic of physically similar systems was a short note in the *Journal of the Washington Academy of the Sciences* (the precursor of *Proceedings of the National Academy of the Sciences*) in July 1914. This first, short piece on the topic is notable for the unprecedented generality of its approach. It did not imply that there were any set fundamental quantities, nor how many there were. It did not talk about physics, even. It talked about quantities, relations between quantities, and equations. All these features of his treatment are sympathetic to the approach urged by Ostwald—a focus on equations rather than models and mechanisms, and a more abstract and open-minded approach about allowable kinds of quantities. (That is, Ostwald thought that energy and energy transformations were the basic concept, and he also thought that there were many different kinds of energies, some yet to be recognized.) But Buckingham's treatment had a logical aspect that went beyond the point about the kinds of quantities allowed. His discussion was completely in terms of quantities, relations, and equations, in the most general sense of the words. Although the treatment in Buckingham's short July 1914 paper addressed equations that involved products and sums, it quickly moved to showing that an even more general form of the equation, sans symbols for arithmetic operations, was all that was needed to establish similarity. It contained no examples of systems or applications, save Buckingham's brief remark in closing: "A particular form of this theorem, known as the principle of 'dynamical similarity,' is in familiar use for the interpretation of experiments on mechanical models; but the theorem is equally applicable to problems in heat and electromagnetics."

The treatment of physically similar systems in his July 1914 paper is spare and elegant; it begins bluntly: "Let n physical quantities, Q, of n different kinds, be so related that the value of any one is fixed by the others. If no further quantity is involved in the phenomenon characterized by the relation, the relation is complete and may be described by an equation of the form $\Sigma \, M \, Q_1^{b1} \, Q_2^{b2} \, Q_3^{b3} \ldots Q_n^{bn} = 0$ in which the coefficients M are dimensionless or pure numbers." (Here Σ is the mathematical symbol indicating a sum of products.) He then lays out the points illustrated in Figure 3. The relation between physical quantities can be put into a more general form: $\varphi \, (\pi_1, \pi_2, \pi_3, \ldots \pi_i) = 0$, where π_is are dimensionless products of the quantities (Qs) and are independent of each other. Since the π_is are dimensionless, their value is independent of the size of the fundamental units used. Notice also that in the "most general form" of an empirical equation he gives, no symbols for any arithmetic operations (such as addition or multiplication) occur.

There is a generality to Buckingham's treatment in that it does not specify which units are fundamental, nor the size of the units to be used for any quantity. In fact, he makes it clear that it is a matter of choice which units are to be regarded as fundamental ones. "Let k be the number of fundamental units needed in an absolute system for measuring the n kinds of quantity. Then among the n units required, there is always at least one set of k which are independent and not derivable from one another, and which might therefore be used as fundamental units, the remaining ($n - k$) being derived from them." Together, these allow him to say how the quantities *other than* those that are taken to be among the k fundamental quantities are related to those k fundamental quantities (however many there are and however they may be chosen). Denoting the fundamental units by $[Q_1]$, $[Q_2]$, and so on up to $[Q_k]$, where the square brackets indicate the units of the enclosed quantity, and the remaining ($n - k$) units that are derived from them by $[P_1]$, $[P_2]$, and so on up to $[P_{n-k}]$, we get ($n - k$) equations that relate the units of the ($n - k$) Ps to the units of the k Qs. It can then be shown that the number of independent π_is is ($n - k$).

It is this very general treatment that marks Buckingham's work; it is even more general than he implies, for it is not even restricted to sciences that have been developed up to the point of knowing the equations that describe the phenomenon of the science. *It is applicable to any phenomenon that one might describe in terms of relations.* In a paper several years later further developing the points in his 1914 papers, he explains:

> Dimensional equations are conventional short-hand descriptions of general relations which subsist among physical quantities of different kinds, and the distinguishing peculiarity of dimensional reasoning is that it uses the known facts of Physics only in this general form. Instead of making detailed quantitative assumptions about the phenomenon under consideration, we make only the qualitative assumption that it may be adequately described in terms of a certain set of quantities [...]

> As regards accuracy, it may be remarked that the results of dimensional reasoning are subject to the same limitations as those of any other theory. Theory always operates on an ideally simplified picture of reality because real phenomena are unmanageably complicated. The results obtained are not *exactly* true for any real phenomenon, though they may be for an ideal one; and the approximation with which a theoretical equation, however obtained, represents the actual facts, always depends on the approximation in *essentials* between the ideal picture and its real prototype.

Continuing with the presentation in Buckingham's July 1914 paper, instead of all the quantities involved being of different kinds, suppose some quantities of the same kind are involved. (He gives no examples, because the paper is a very abstract treatment, but to help understand what he means, an example here might be a geometric ratio, such as an aspect ratio used to study wing design, which is a ratio of a length to a height.) If we introduce these pure ratios (ratios of quantities of the same kind) "in an entirely general way by introducing them as additional arguments of the function φ," we obtain an equation of the form $\varphi\,(\pi_1, \pi_2, \pi_3, \ldots \pi_{n-k}, r', r'', r''' \ldots$ etc.$) = 0$, where k is the number of fundamental units. The effect of these pure ratios can then be included when we determine the consequences of the principle of dimensional homogeneity, which imposes limitations on the forms of physical equations.

Buckingham states the principle of dimensional homogeneity as follows:

> Any complete physical equation which describes a relation subsisting among quantities of n different kinds, of which k kinds are independent and not derivable from one another, is reducible to the form

$$\Psi\,(\pi_1, \pi_2, \pi_3, \ldots \pi_{n-k}, \mathrm{r'}, \mathrm{r''}, \mathrm{r'''} \ldots \text{etc.}) = 0 \qquad (*)$$

> in which the r's are all the independent ratios of quantities of the same kind, and each π is determinable from an equation of the form

$$[\pi] = [Q_1{}^\alpha, Q_2{}^\beta \ldots QK^k] = [1] \qquad (**)$$

We have seen the bracket notation before, in Nichols' lecture on dimensional formula; brackets around a variable indicate the dimensions of that variable. Hence, $[v]$, where v is a velocity, indicates dimensions of (length)(time)$^{-1}$. The dimension of any pure number is 0, so $[1]$ is used to indicate a dimension of 0. That $[\pi] = [1]$ indicates that π has dimensions of 0, which is to say that it is a dimensionless parameter.

This is a very general statement of the principle underlying the use of dimensional reasoning that occurs so often in the many discussions on scaling and similarity that appeared around 1913 to 1914, and it is more general than anything discussed in this book so far. Inasmuch as it was invoked, it was usually implicit and was seldom justified or explicitly stated. Generalizing even further, he shows that, if h is the whole number of quantities, "whether of different kinds or not," "any complete physical equation is reducible to the form $\Psi\,(X_1, X_2, \ldots X_{h-k})\,[= 0]$, in which the Xs are all the possible independent dimensionless combinations of all the quantities involved." In later papers, he was not as keen to show the most general representation in symbolic logic he could attain. The goal here, in his very first and very lean paper with no examples or applications, was to get straight on things that had not yet been articulated by others who had employed the method. He would later write to Rayleigh about these early papers:

> I had therefore, as it were, to write an elementary textbook on the subject for my own information. My object has been to reduce the

method to a mere algebraic routine of general applicability, making it clear that Physics came in only at the start in deciding what variables should be considered, and that the rest was a necessary consequence of the physical knowledge used at the beginning; thus distinguishing sharply between what was assumed, either hypothetically or from observation, and what was mere logic and therefore certain.

The resulting exposition is naturally, in its general form, very cumbersome in appearance, and a large number of problems can be handled vastly more simply without dragging in so much mathematical machinery.

The point on which our story turns occurs in the observation Buckingham makes soon after presenting the very general form to which any physical equation can be reduced: that "The chief value of the principle of dimensional homogeneity is found in its application to problems in which it is possible to arrange matters so that the r's and Πs of equation [(*)] shall remain constant and the definite equation [(**)] therefore be satisfied," where equation (*) is an equation solving for a particular quantity P1, and equation (**) is the special case of that equation when all the r's are held constant and the variations of all the quantities (all the Ps and Qs) are varied, in such a manner that all the Πs remain constant.

The consequence of this principle that he draws for similar systems is illustrated in Figure 3. (The figure is not taken from Buckingham's paper; it is my own depiction of the points made in his paper.) The consequences for similar systems are extremely interesting, for although the paper begins with a discussion of how equations depict relations that hold in the world, the consequences Buckingham draws concern how one can construct a system in which a measurement in the constructed system allows one to determine the value of a quantity in the other system to which it is similar. What is striking is that the constraints that must be satisfied in constructing this system are the value of the dimensionless parameters that appear in the most general form of the equation that depicts relations that hold in the world.

For someone who had been immersed in the discussions about models and equations in Leipzig and Vienna in the time of Boltzmann and Ostwald, here was the sought-for link! That is, debates there about science often pointed out that there were two different ways to depict something in physics: by equations, and in terms of models (either imagined, as in a thought experiment, or actual, as in constructed scale models). Which of these was appropriate and informative in the kinetic theory of gases was, of course, at issue in debates between Buckingham's doctoral director, Wilhelm Ostwald, and Ludwig Boltzmann, with Ostwald deriding the use of models as unsound reasoning and emphasizing equations, and Boltzmann defending the use of models in science, with Boltzmann making the interesting observation that manipulating equations is somewhat like manipulating models. Now we see that in the explicit statement of the principle of dimensional homogeneity there is a link between them: *the model is constructed in a way that is guided by the equation*, and we can state precisely *how* it is guided by the equation. Furthermore, the rules of correspondence between corresponding quantities in similar systems are deducible from the equation. These, however, are not meaningful statements of science, or about how things are in the world; they are the practical rules of how to "hold the model up to the world," so to speak.

It is not crucial that we be able to state the equation in its most general form; *what is crucial is to identify a set of independent dimensionless groups*. That is, in order to pick out the dimensionless groups that guide model-construction, we do need to be able to see at least this much of its underlying form. The most general form of the equation as Buckingham gives it is very like the form of equations used in Lagrangian mechanics: there, the governing equations are stated in terms of generalized coordinates and some constraints. The generalized coordinates are all independent of each other, and there are as many of them as there are degrees of freedom in the system. Rather than using force balances, as one would when using Newton's equations of motion, a single principle is used—"the principle of least action." Applying this principle effects (brings about) the solution of the equations, and thus provides an alternative to working out the detailed equations of motion one often gets when applying Newton's laws. The values of the familiar Cartesian coordinates can be found by transformations, and there are as many equations as there are degrees of freedom of the system. Buckingham's paper follows along the lines of laying out analogues of all the things just mentioned, but his subject is not limited to the equations of mechanics: it applies to any empirical equation whatsoever. The analogue of the principle of least action in his paper is a principle of *logic* rather than a principle of physics: the requirement of dimensional homogeneity. The method underlying the construction of physically similar systems is not a method peculiar to mechanics; it applies to *any equation describing a complete relation that holds between quantities.*

In providing a more general formulation of the equation in terms of dimensionless groups of the quantities involved, Buckingham is not implying that there is anything *wrong* with the equation as stated in the language commonly used in a discipline. It is just that it can be very useful to discern the more general form, if one is interested in clarifying how many independent dimensionless parameters (or "groups") there are and wants to identify a set of independent dimensionless groups for guidance as to how to vary quantities in a system in order to transform it into a system that behaves similarly. Thus, providing the most general form of the equation was not an effort to *reform the practice* of using the kind of equations then currently in use; it was pointing out that the more general form of the equation perspicuously shows the constraints needed for model construction; it pares down the equation to only what is relevant to it from the standpoint of the method of dimensions.

Therein lies the relevance of the presentation Buckingham gave in his 1914 papers to our story. There really was nothing like it in the English-speaking world prior to his 1914 papers, and probably not in the German-speaking world, either. There was something nearer to it, though still not at the same level of generality, in Russia and France: a paper authored by the scientist associated with Russia's wind tunnel, Riabouchinsky. Some instead credit Vaschy with the first proof of the theorem. After Buckingham became aware of Riabouchinsky's work, he cited Riabouchinsky as the author of the proof involved, though he credits Rayleigh for the method of dimensions in general. In a later 1921 paper published in a British journal, he writes:

> The dimensional method of analysing problems in Physics, which we owe to the late Lord Rayleigh, somewhat resembles the methods of Thermodynamics in being very simple in principle but sometimes a little puzzling in practice. To the initiated, all the information obtainable from dimensional considerations is often evident upon mere inspection, so that any formal and detailed account of the reasoning appears quite superfluous; and Lord Rayleigh's numerous applications are sometimes so concisely described that the results seem

rather like magic. But while a closer study of his solutions can only increase our admiration, it will certainly lead the average reader to wish for a less intuitive and more systematic procedure for obtaining the same result.

Such a routine procedure is provided by formulating the requirement of dimensional homogeneity as a general algebraic theorem, which was first published by Riabouchinski (sic), and which will be referred to as the π theorem.

Concerning Riabouchinsky's work, he cites the 1911 paper in *L'Aerophile* and a 1912 paper in the *Koutchino Bulletin*, explaining:

A reference to the first of these papers [by Riabouchinski (sic)] appeared in the *Annual Report of the British Advisory Committee for Aeronautics* for 1911-1912, p. 260, abstract 134. Guided, no doubt, by the hint contained in this abstract, the present writer came upon substantially the same theorem and described it, with illustrative examples, in the "Physical Review" for October 1914 (vol. iv, p. 345). The statement of the theorem given in the present paper does not differ materially from Riabouchinski's, except in that he confined his attention to mechanical quantities.

The theorem in its more restricted form as presented by Riabouchinsky—that is, as a theorem about equations involving mechanical quantities—would have been available in periodicals devoted to aeronautics just as Wittgenstein, though still registered as a research student in engineering at Manchester, went to Cambridge in autumn of 1911 to study logic with Russell. As mentioned earlier, according to Russell, Wittgenstein often regarded himself as "an aviator" and still considered himself undecided about whether to be in aeronautics or philosophy during his first term at Cambridge. So it is not implausible that Wittgenstein had occasion to read or discuss what was in aeronautical periodicals. He could have been aware of the less-general theorem through the appearance of Riabouchinsky's work, either through the British Advisory Committee on Aeronautics (ACA) report for 1911–1912 or through the 1911 article in the French magazine *L'Aerophile*. If so, some of the general ideas (though not their coalescence into an explicit statement) about there being something exhibited in the most general form of an equation of mechanics that serves also to establish the conditions for dynamical similarity of mechanical systems might have already been vaguely in mind during his study of logic with Russell. Then, Buckingham's 1914 articles, setting the problem more generally in terms of quantities of *any kind related by some relation R,* would have amounted to a generalization of something he already knew to hold in mechanics—and, significantly, a generalization to the more general case of relations in the world described by scientific equations: that they have an underlying, more general and perspicuous form that can be written in terms of symbolic logic. Buckingham's approach, too, was much more self-aware of the role of symbolic notation in his analysis; he often mentioned the activity of representing relations by equations and thus invited questions in logic and philosophy.

In the paper that appeared in October 1914 (in *Physical Review,* responding to Richard Tolman), the approach is more marked in two respects. The paper begins right off with the subject of representing relations by symbolic means. The very first paragraph is subtitled "The Most General Form of Physical Equations," and the first topic raised is the task of representing: "Let it be required to describe by an equation, ..." Secondly, he mentions that the function to which the equation can be reduced involves a further choice: "when we have found the specific form of any one of the Πs [independent dimensionless parameters], we are at liberty to replace this by any function of it; for this function will also be dimensionless and will be independent of the remaining Πs."

At any rate, in his first paper on physically similar systems (the July 1914 paper), Buckingham went on to explain in more detail what he meant in saying that the chief value of the principle of dimensional homogeneity lay in its applications to problems for which it was possible to arrange things so that the dimensionless ratios were held constant. It is at this point that he introduced the idea of a transformation of one system into another by varying some of its quantities in an orchestrated manner. Transforming a screw propeller or a wing from a small to a larger version is one example of such a transformation, but he emphasized that it worked for transforming electrical and magnetic laboratory setups and devices as well:

> The quantities involved in a physical relation pertain to some particular physical system which may usually be treated as of very limited extent. Let S be such a system, and [let us denote by (*)] the equation

which describes a relation subsisting among certain quantities of the kinds Q and P which pertain to S, e.g., the sizes, densities, thermal conductivities, etc. of its essential parts.

Let S' be a second system into which S would be transformed if all quantities of each kind Q involved in equation (*), were changed in some arbitrary ratio, so that the r's for all quantities of these kinds remained constant, while the particular quantities Q_1, Q_2, ... Q_k changed in k independent ratios. For example, if Q_1 is a length, S and S' are to be geometrically similar in all their essential parts, though other parts, of which the size and shape have no influence on the relation under consideration, are not subject to any geometrical conditions. The systems now "correspond" as regards their essential parts and may be said to be similar as regards each of the kinds of quantity Q separately, so far as such quantities are involved in equation (*).

The Qs are the fundamental quantities. However, recall that there is some leeway as to which quantities are considered fundamental and which are derived. The ones that were not used as fundamental were designated as Ps. Hence, the derived quantities P and all quantities of that kind involved in the relation need to be transformed as well, in such a way that the Πs remain constant. Then, "[t]wo systems S and S' which are related in the manner just described are similar as regards the physical relation in question." These are criteria that would need to be met for similarity; he hints that it may not be possible to meet all the criteria for similarity.

It's rather obvious that the notion of similar systems—one system S being transformed into another system S' in such a way that it "corresponds" to S ("as regards the essential quantities")—is relevant to evaluating Tolman's claim that the universe could be transformed overnight into an observationally indistinguishable miniature universe (in his 1914 "Principle of Similitude" paper in *Physical Review*). It is also relevant to Stanton and Pannell's "similar motions" paper, in that it is a more general treatment of the methodology of model testing in wind tunnels ("the principle of dynamical similarity") given there. And, as a matter of fact, in the next paper Buckingham wrote (which also came out in *Physical Review*), in addition to presenting the generalized treatment found in the July 1914 version of "Physically Similar Systems," he addressed both these related topics on which major papers appeared in early 1914: he discussed experimental engineering model testing, and he replied to Tolman's claims about the possibility of an observationally indistinguishable miniature universe.

The illustration of experimental scale modeling he uses is from naval research: the screw propeller. His presentation in the later 1914 paper follows the logic laid out in the July 1914 paper. Though the relation describing the behavior of a screw propeller is unknown, it is supposed that, whatever that relation is, it involves seven quantities of seven different kinds: force F, density ρ, length D, linear speed S, revolutions per unit time n, viscosity υ, and acceleration g. Also, it is understood that three fundamental units are needed. Various sets would do, such as {mass m, distance l, time t} or {force F, density ρ, linear speed S}. If the {m, l, t} set is used, F has units of $[mlt^{-2}]$, ρ has units of $[ml^{-3}]$, D has units of $[l]$, S has units of $[lt^{-1}]$, n has units of $[t^{-1}]$, υ has units of $[ml^{-1} t^{-1}]$, and g has units of $[lt^{-2}]$.

On the supposition that seven kinds of quantities are involved in the equation that describes the relation that determines the phenomenon of interest associated with the screw propeller, and three fundamental units, the equation is reducible to an equation in terms of an unspecified function of four (7 − 3) independent dimensionless parameters. Thus, it can be reduced to the form

$$\Psi\,(\Pi_1, \Pi_2, \Pi_3, \Pi_4) = 0$$

which is the form that Buckingham called "the most general form of a physical equation." To find the dimensionless parameters Π_1, Π_2, Π_3, and Π_4, three of the seven quantities are denoted as fundamental. Some freedom of choice exists; he chooses force F, density ρ, and length D. These are denoted as Q_1, Q_2, and Q_3; the remaining four quantities are denoted by P_1, P_2, P_3, and P_4.

Buckingham's use of exponents is as interestingly different from the usual interpretation as it is potentially confusing. Considering all four possible products consisting of one of the Ps along with the three fundamental quantities, it is asked what exponents each quantity (Qs and Ps) could have such that each of the four possible products would be dimensionless. He explains in his later 1914 paper that the exponents indicated *repeated operations*. Wittgenstein's use of operations in the *Tractatus* has sometimes dismayed and puzzled its readers. Could this notion of operations help clarify his intent?

The reason why such an expression as Q^2 can appear is that Q^2 may be regarded as a symbol for the result of operating on Q by Q. For example, when we write $A = l^2$, l^2 is a symbol for the result of operating on a length l by itself. We are directed to take the length l as the operand and "operate on it with the length l" by constructing on it as a base, a rectangle of altitude l; the result of this operation, which fixes an area A, is represented by l^2.

Thus, integers are part of the symbolism, but they appear *only* as exponents that indicate repeated operations. This is exactly the role to which Wittgenstein relegated integers in the *Tractatus*!

The Wittgenstein scholar Juliet Floyd has delved into the question of where Wittgenstein's use of operations might have come from. She draws attention to the work of Whitehead, who writes of operations: "The derivation of a thing p from things a, b, c, . . . , can also be conceived as an operation on the things a, b, c, . . . , which produces the thing p." It's possible that this account of an operation fits with the kind Buckingham had in mind. After all, both Buckingham and Whitehead come from a milieu of late nineteenth-century logic and philosophy of science. She also notes that the account of numbers as exponents found in Wittgenstein is distinctively different from accounts of number in Russell and Frege, and she considers his statements about basing arithmetic proofs on this account of numbers.

These exponents are central to practical uses of dimensional analysis. Applying the principle of dimensional homogeneity yields a set of simultaneous linear equations in which these exponents are the unknowns. Using simple linear algebra to solve for the exponents, four dimensionless groups—Π_1, Π_2, Π_3, and Π_4—are obtained. The algebra and arithmetic that are involved in the process of determining the dimensionless groups arise from the notion of a (repeatable) operation. However, algebra and arithmetic are not part of the symbolism used to describe the relations in the world. The algebraic equations and thus the sign for addition arise only from our explicitly stating *a logical constraint on* empirical equations, not in the general form of the empirical equation itself.

As mentioned earlier, in dimensional equations a dimensionless parameter can be replaced by a suitable function of it , which permits elimination of fractional exponents. For the screw propeller case and the choices made for Qs and Ps earlier, the following set of dimensionless Πs results after the fractional exponents are eliminated: $\Pi_1 = (\rho D^2 S^2/F)$, $\Pi_2 = (\rho Dn^2/F)$, $\Pi_3 = (\mu^2/F\rho)$, and $\Pi_4 = (\rho D^3 g/F)$. Since the assumptions made are not specific to the screw propeller, it is possible to conclude "that any equation which is the correct and complete expression of a physical relation subsisting among seven quantities of the kinds mentioned is reducible to the form $\Psi ((\rho D^2 S^2/F), (\rho Dn^2/F), (\mu^2/F\rho), (\rho D^3 g/F)) = 0$." However, other values of the dimensionless parameters Πs exist, so other equations of the same form are possible. For example, choosing Q_1, Q_2, and Q_3 to be ρ, D, and S, respectively, yields $\Psi ((\rho D^2 S^2/F), (Dn/S), (\rho DS/\mu), (Dg/S^2)) = 0$. Including the fixed pure ratios (r', r'', etc.) of the problem as well, the most general form is:

$$\Psi ((\rho D^2 S^2/F), (Dn/S), (\rho DS/\mu), (Dg/S^2), r', r'', \ldots) = 0$$

There are no signs for any algebraic operations in it! The function is not even determined. All the equation indicates is that a function exists—an undetermined one. There is practically no structure or form to this equation at all, at least on the surface. All the structure of such an equation has to come from the system of symbolism itself.

Buckingham then shows how this equation, though it is in terms of an undetermined function, informs us how to construct a model from which the thrust of a screw propeller of a larger size can be determined experimentally. If any three of the dimensionless parameters are set, the fourth is determined. Though he does not mention it, this way of putting it is reminiscent of van der Waals' law of corresponding states emphasized by Onnes in his 1913 Nobel lecture ("The reduced temperatures of two fluids are the same if the fluids are in corresponding states, that is, at the same reduced volume and pressure."). We can rewrite the general form of the equation (which is written in terms of an undetermined function) as an equation relating one of the dimensionless parameters to the other three by a (different) undetermined function. If F occurs only in that dimensionless parameter, as it does in the preceding equation, it is possible to solve for F in terms of another undetermined function of the other three dimensionless parameters, eventually obtaining an expression for F as follows:

$$F = \rho D^2 S^2 \, \varphi ((Dn/S), (\rho DS/\mu), (Dg/S^2), r', r'', \ldots)$$

This then tells us how to construct an experimental model propeller and how to draw inferences about the full-size propeller from it.

The principle of dynamical similarity states that in passing from one screw propeller to a second, in the same or in another liquid, any three kinds of quantity, such as (ρ, D, S), that can provide fundamental units, may be changed in any ratios whatever, and that the equation that connects the thrust with the other quantities remains precisely the same if the values of the arguments of ϕ remain unchanged. This means, in simpler language, that:

If we find the value of the constant N in the equation

$$F = N \rho D^2 S^2$$

from an experiment in which the arguments of ϕ have a certain fixed set of values, the same constant is applicable to any values of (ρ, D, S) if the values of (Dn/S), (ρDS/μ), (Dg/S^2) and the r's are the same in the second case as in the first.

The simplest of the requirements for the useful application of [F = ρD^{2}S^2 ϕ ((Dn/S), (ρDS/μ), (Dg/S^2), r', r'', ...)] is that the r's shall be constant; hence the two propellers, whatever their diameters, must be geometrically similar and similarly immersed; and the smaller may be called the model while the larger is called the original. The next simplest condition is that Dn/S shall remain constant. . . . The blades being of a fixed shape, the condition that Dn/S shall be constant is the same as the condition that the "angle of attack" of the blades on the still water into which they are advancing shall be constant. . . . Our two conditions may . . . be expressed by saying that for two screw propellers to be dynamically similar, they must first of all have the same shape and be run at the same relative immersion and at the same slip ratio.

So some practical reasoning is involved in laying out the conditions for dynamical similarity. However, this is only the beginning of the process of model building. In examining how one would go about actually varying all the quantities in concert so as to keep the needed dimensionless Π's constant, Buckingham concludes that in fact "we can not, in practice, run a reduced-scale model screw propeller so that it shall be dynamically similar to its original. We must therefore limit ourselves to a less ambitious program and attempt to obtain an approximate result . . .; and to do this we must find a plausible pretext for omitting one of the two arguments of φ." This is not an insurmountable obstacle. "For it is apparent from various hydrodynamic experiments that when a fluid is in very turbulent motion its mechanical behavior is little influenced by viscosity, density being much more important." After developing conditions for dynamically similar propellers, Buckingham then remarks on what is to be learned from the illustration:

> By disregarding viscosity we have, in effect, disregarded the effect of skin friction on the action of the propeller; and we have also left aside the question of cavitation. But without venturing further into the chaos of screw-propeller theory, the foregoing example will serve to illustrate the sort of use that may be made of dimensional reasoning in attacking mechanical problems which are—like most of those that occur in practical hydro- and aerodynamics—too difficult to be handled at all by ordinary methods.

His detailed analysis of how the notion of similar systems—especially of transforming one system into another—is involved, what its shortcomings are, and how those shortcomings can be compensated for with experimentally obtained knowledge of the particular kind of problem being investigated, thus providing a high level of accuracy in spite of the approximations made, explains both why it was possible to get such good results and why getting good results was, after all, partly dependent on the art of an expert modeler in the field. What Buckingham showed here was that the fact that getting good results was so dependent on the art of the modeler does *not* mean that the method was not amenable to formalization, or that it was not informed by sound mathematical and physical principles.

Being a physicist, and one interested in the foundations of thermodynamics at that, Buckingham's interest in the notion of similar systems extended beyond showing its role in the method underlying the use of experimental scale models, as thorough as the account of that practice he gave in the 1914 *Physical Review* paper may be. In the same paper, he goes on to promote "the more general conception of a similarity which extends to other than merely dynamical relations" which, he observes, appears to follow from dimensional reasoning based on "the principle of homogeneity."

Actually, in the 1914 *Physical Review* paper, he says that his whole purpose in illustrating the principle of physical similarity with the detailed exposition of how it is employed in experimental modeling of the screw propeller was to provide background against which to respond to Tolman's proposed "principle of similitude." After remarking that the notion of similar systems used in constructing and using a model propeller is generalizable beyond mechanics, he shows how the principle involved in doing so—the "method of dimensions"—applies in problems ranging from electrodynamics (energy density of a field, the relation between mass and radius of an electron, radiation from an accelerated electron) to thermal transmission, and, finally, at a higher level, to the kind of "bird's-eye view" question to which his interest tended to migrate: "the relation of the law of gravitation to our ordinary system of mechanical units." He had just several years earlier written on the relation of Newton's first law to the unit of time, and he had been critical of "nonsense about clocks 'running at a constant rate,'" declaring that the term "constant rate" was "quite meaningless except in connection with some time scale outside of and independent of the clock itself."

The question he asks about the role of the law of gravitation in determining units of measure is a bit different. It is about the number of "fundamental units," and the question Buckingham asks can be put in terms of similar systems: if it is in fact true that in mechanics three fundamental units suffice to describe mechanical phenomena (more if thermal and electromagnetic phenomena are to be described), then it would be correct to conclude that

> a purely mechanical system may be kept similar to itself when any three independent kinds of mechanical quantity pertaining to it are varied in arbitrary ratios, by simultaneously changing the remaining kinds of quantity in ratios specified by [the equation $[\pi] = [Q_1{}^\alpha Q_2{}^\beta Q_3{}^\gamma P] = [1]]$. . . . For instance, we derive a unit of force from independent units of mass, length, and time, by using these units in a certain way which is fixed by definition, and we thereby determine a definite force which is reproducible and may be used as a unit. Now by Newton's law of gravitation it is, in principle, possible to derive one of the three fundamental units of mechanics from the other two.

Buckingham then describes a laboratory experiment from which a unit of time can be derived from units of mass and length. The setup involves two free masses allowed to approach each other; the resulting time interval is a function of the masses and the distance between them. The point is not about the unit of time per se, but about the reduction in the number of units required in mechanics: "if we utilize the law of gravitation, only two fundamental units are needed for mechanical quantities, instead of the three that physicists ordinarily use." This might make little difference to whether or not we actually use different units for mass, length, and time, but even so, there is still the question of the how many parameters can be varied in the process of constructing physically similar systems.

Recall that, in constructing one system to be similar to another, the criterion for physical similarity (with respect to the relations among a certain number of kinds of quantities) was that "certain of the kinds—equal in number to the fundamental units required for the absolute measurement of all the quantities involved in the relation—are subject to variation in arbitrary ratios, if we fix the ratios in which the remaining kinds of quantity shall then change by imposing the condition that the Πs shall remain invariable." So it's really of crucial and practical importance that we know how many fundamental units are required to describe the relation of interest for the physical system under consideration. Otherwise, we are *not* free to vary three quantities in transforming a physical system into one that is physically similar to it, and the treatment of the screw propeller problem is flawed. Which is it, then? Two or three fundamental units?

Surprisingly, Buckingham's answer is that, even within the domain of mechanics, it depends. It depends on what phenomenon the relation characterizes. The notion of physical similarity and physically similar systems devolves around a specified relation. Recall that the analysis started with the quantities involved in a given equation, where that equation describes a relation relating a certain number of kinds of quantities such that any one was determined by all the others, and the relation characterized a phenomenon of interest. Although he spoke of systems rather than of equations, the process of transforming one system into another to which it was physically similar also was relative to a relation relating a certain list of kinds of quantities, and such a relation is often expressed in an equation.

In developing a general methodology, Buckingham had considered *any* such relation—that is, all *possible* relations that could exist among the given kinds of quantities. He concludes that if, on the contrary, we consider only *some* such relations, we can take advantage of some features of specific relations:

> In the most general case, when we include within the field of our reasoning all kinds of physical quantities and all possible relations among them, we must admit our familiarity with the law of gravitation and limit ourselves to two fundamental units. But if for "all possible relations" we substitute "all relations that do not involve the law of gravitation," we may ignore the law and proceed as if it were nonexistent.
>
> With this single proviso all our foregoing reasoning retains its full validity. The limitation is seldom felt, because, in practice, physicists are seldom concerned with the law of gravitation: for all our ordinary physical phenomena occur subject to the attraction of an earth of constant mass and most of them occur under such circumstances that the variation of gravity with height is of no sensible importance.

The situation is different for some problems, though. In "precise geodesy" and in astronomy, the complete description of the phenomena being investigated *does* need to make explicit that the law of gravitation is operating. Then, "If the physical relations which characterize such phenomena are under discussion, we *must* recognize the law of gravitation, we *must* regard all mechanical units as derivable from two and not three independent fundamental units, and if a physical system is to remain similar to itself . . . if we exclude thermal and electromagnetic quantities, only two [arbitrary changes are possible.]" (In a later letter to Rayleigh, he admitted it was a mistake to say "must" here: "in my *Physical Review* paper, I fell to the error to which you called my attention . . . I said, in effect, that we must use the equation for the law of gravitation at the beginning of our algebraic process, and that is obviously not necessary but only permissible, as I saw at once upon further consideration of the problem of the liquid globe.")

Buckingham's response to the question of the possibility of constructing observationally indistinguishable miniature universes bifurcates into two cases, depending on whether or not the phenomenon that we are interested in observing in the miniature universe is influenced by the law of gravitation. If not, then it is possible to construct a miniature universe as Tolman suggests that will be similar to the universe (as regards that phenomenon). On the other hand, if the phenomenon is influenced by the law of gravitation, more things must be taken into account: "the gravitational forces in the miniature universe must bear to the corresponding gravitational forces in the actual universe a ratio fixed by the law of gravitation." He points out that the effect of the law of gravitation on the phenomena of interest shows up in the process of constructing similar systems. If we erroneously try to independently choose three units rather than letting the third be determined by the first two fundamental units chosen, we run into trouble because the measured values for corresponding speeds and forces won't correspond to the values in the actual universe—unless, that is, the third unit is allowed to be fixed by the law of gravitation in terms of the first two.

In later correspondence with Rayleigh, Buckingham tried to explain his view another way: one could regard the gravitational constant as a universal constant—that is, as something that "should not be treated as a quantity which could ever or *under any circumstances*, be varied." In this case, there are only two fundamental units, for the unit of mass can be replaced by units of l^3/t^2. Alternatively:

> if [the gravitational constant] is a *quantity*, the law of gravitation does not enable us to eliminate one fundamental unit. Hence we must retain our three fundamental units and also treat [the gravitational constant] as a quantity which happens, to be sure, to be constant but which might conceivably have varied. . . . I believe you are right in suggesting that the difference is mainly one of form;

The points about physically similar systems, systems of units, and the law of gravitation seem to be questions in the logic of physics. So it is a little surprising to realize that the main claim of Buckingham's papers on physically similar systems can actually be stated in terms of a theorem about the *symbolism of relations* between physical quantities. The paper is summarized in terms of the form to which a complete physical equation can be reduced. Examining the wording closely, it is actually a statement about symbolizing relations between kinds of physical quantities. *The equation is involved only insofar as it picks out a relation* between quantities of certain kinds. The most general form of the equation *is not obtained by transformations on the equation*. In fact, it cannot be, because the most general form of the equation, $\varphi\,(\pi_1, \pi_2, \pi_3, \ldots \pi_i) = 0$, is obtained by dimensional considerations concerning how the different kinds of quantities related (by the relation picked out by the equation) can be combined to form dimensionless products. Hence, it is a symbolism that reflects the logical relations between the quantities, not the picked-out relation itself. Thus, it is fair to regard finding the most general form of the equation as finding the underlying logic of the equation; it is not equivalent to the equation in cognitive value, in that it is stated in terms of an unspecified function φ.

This is the same sort of situation Frege found with respect to sentences of natural language and the thoughts they express: the sentences of a natural language often express more than the thought. The formulae of the *Begriffschrifft* were meant to capture only the thought—what was important to making inferences. Likewise, the most general form of an equation here is not meant to provide an alternative form of the equation for the same uses in scientific practice that the equation may have. Its value is in illuminating the logical features of the relation picked out by the equation, and those are captured in a (not necessarily unique) set of dimensionless parameters. What is unique is the minimum number of dimensionless parameters needed. Knowing how many independent dimensionless parameters there are has a practical significance in the design of model experiments, and determining that number is the motivation for, and the main value of, the theorem he presents, but Buckingham does not even mention the number of independent parameters in his closing summary:

> A convenient summary of the general consequences of the principle of dimensional homogeneity consists in the statement that any equation which describes completely a relation subsisting among a number of physical quantities of an equal or smaller number of different kinds, is reducible to the form $\varphi\,(\pi_1, \pi_2, \pi_3, \ldots \text{etc.}) = 0$ in which the π_i's are all the independent dimensionless products of the form $Q_1{}^x$, $Q_2{}^y, \ldots$, etc. that can be made by using the symbols of all the quantities Q.

That the consequences of the principle provide a theorem about *symbols for relations of quantities* of different kinds is striking.

The significance of the theorem can be seen in the sketch in Figure 3: if the symbolism for the actual relationship in the world (expressed by a complete physical equation) is properly discerned, it quite directly tells us things that must hold for any system that is physically similar to the one in which the relation described by the scientific equation holds. The scientific equation might be thought of as a sign, in contrast to the symbolism exhibited by the most general form of the equation, for the latter not only mirrors the relation between quantities of different kinds but also shows what physical similarity with respect to that relation consists in. The general form to which the scientific equation is reducible tells us how many degrees of freedom there are—that is, how many quantities may be arbitrarily varied (while the rest are unchanged). This same symbolic form—the most general form of a complete physical equation—tells us how to map quantities in the transformed or model system to corresponding quantities in the actual system.

I have tried to lay out these points in Figure 3 so as to highlight that it is quite natural to think of the scientific equation and the similar system or model system having some logical structure or form in common in the same way that the musical notation of a score and the lines on a gramophone record might be said to have something in common—and to think of that logical structure or form as explaining how it is that the equation can describe the actual system and the model can be physically similar to the actual system. Loosely put, the explanation is that they share a common logical structure, but for the model the structure is a matter of criteria for similarity to systems in the world with respect to a certain relation between quantities of different kinds. Because the form—the most general form of a physical equation—is involved in several distinct ways, it is helpful to reflect on Figure 3, which shows how the dimensionless parameters figure in explaining these several different ways.

The year 1914 thus marked not only an abundance of interest in mechanical and dynamical similarity in many different fields and, correspondingly, attempts to explicitly state bases for it, but also an unprecedented generalization to a general notion of similarity. The general principle was stated as a principle of logic, a principle about the logical form of empirical statements of relations holding between quantities of different kinds. What mattered was the number of *kinds* of quantities involved in the relation, and the number of fundamental units that were relevant in describing the phenomenon of interest determined by the equation describing that relation.

Buckingham was aware that, stated in it most abstract form, the theorem might look "rather noncommittal." But he was convinced of its importance—and of its potential. As someone who had seen the power of energy methods in elegantly solving otherwise-tedious problems (such as the use of the energy principles in mechanics), and who was cognizant of the far-reaching consequences of energy principles that look on the surface as though they say little, he could appreciate its potential. The statement about the most general form to which any physical equation is reducible is, he wrote, "a powerful tool and comparable, in this regard, to the methods of thermodynamics or Lagrange's method of generalized coordinates." As for his purpose in writing about such general principles, he said the paper will have achieved its object if it "merely helps a little toward dispelling the metaphysical fog that seems to be engulfing us."

Although Buckingham's immediate occasion for studying the theory of dimensions seems to have been the initiatives begun in 1911 to build a U.S. capability in aeronautical research, especially model experimentation, the real significance for our story was that, being Buckingham, his inquiry led him far beyond the foundations needed to clarify that practice, all the way to the nature of symbolism. The foundational themes in that paper are not really needed for the application of the method.

In particular, his tastes for foundational questions led Buckingham to delve into the nature of symbolism for expressing relations between different kinds of quantities. Importantly, he distinguished between the signs (equations) used in scientific practice and the symbols used for quantities and their relations in his logical analysis. This could hardly be lost on someone who admired Frege and was struggling to find a proper theory of symbolism. In December 1912, Wittgenstein wrote to Russell from Vienna that he had just had a long discussion with Frege about "our theory of symbolism." The next letter, in January, quoted earlier, shows him still searching for a "proper theory of symbolism," and he writes of his conviction that "every theory of types must be rendered superfluous by a proper theory of symbolism . . . What I am most certain of is not however the correctness of my present way of analysis, but of the fact that all theory of types must be done away with by a theory of symbolism showing that what seem to be *different kinds of things* are symbolized by different kinds of symbols which *cannot* possibly be substituted in one another's places." So, had Wittgenstein heard about Buckingham's paper, which addressed model experiments and a proposed new principle of relativity, and concluded with a theory of dimensions stated in terms of the symbolism used to represent different kinds of measurable quantities, it would certainly have commanded his attention.

Whether any historical evidence (in terms of notes, letters, and so on) exists that Wittgenstein read this paper is an open question. Evidence for anything from this period of his life is scant; he ordered much of it destroyed. There is no reason to rule it out or even to regard it as implausible. He did correspond with William Eccles, the engineering friend he had made at the kite-flying station in northern England, and he made trips to Manchester during this time, visiting Eccles in 1913. He and Eccles often discussed topics in engineering, so Wittgenstein may well have still read or heard about papers on engineering topics. He and Eccles had worked together designing, building, and flying kites, which depended on data collected by whirling arm or wind tunnel, so similarity of physical systems might well have been a topic of interest to them. In July 1914, Wittgenstein wrote to Eccles that he planned to visit him in Manchester in August 1914, although that visit was preempted by his wartime service. Thus, Wittgenstein was in Vienna in July 1914, where foundations of physics was such a lively topic and where access to international scientific literature was easy and common.

Since Buckingham's papers appeared in physics journals, the topic could have arisen as a topic of interest in physics rather than in engineering, either during the summer of 1914 in Vienna or through friends in Manchester. Manchester, too, had a robust German community of scientists and engineers, so Buckingham, as a student of the very famous German professor Ostwald, might have been known among them and a paper by him on physically similar systems noticed. Wittgenstein had spent three years in that community. In general, it is very hard to pin down specific details such as the exact scientific papers he might have read at the time, especially because most of his personal notes and manuscripts from this period were destroyed, on his instructions.

Rayleigh seems not to have known of Buckingham's paper when he wrote his 1915 *Nature* article on "The Principle of Dynamical Similarity," for he wrote to Buckingham in a postscript in a letter to him: "I have forgotten to add that I should of course have referred to your paper had I known of it." This was appended to a letter in which Rayleigh praised the "Physically Similar Systems" paper he had received from Buckingham: "It is the best exposition of the subject I have come across; the only doubt is whether it is not too good for a majority of physicists!" However, even if Rayleigh's comment does reflect that Buckingham's 1914 papers were not well-known among physicists in Britain by 1915, they might well have been known by some physicists in Vienna, especially as that is where there would have been intellectual communities keeping abreast of the activity of Ostwald's American students. In light of Prandtl's comments on how much more appreciated Lanchester's work was in Europe than in his own England, this is more than plausible.

Wittgenstein spent July of 1914 in Austria visiting family and friends, when Buckingham's shorter paper on physically similar systems appeared. Both Buckingham and Tolman, to whom he responded in the later paper of October, wrote philosophical papers on the foundations of thermodynamics, the topic on which Boltzmann, whom Wittgenstein revered, had given outrageously popular lectures in Vienna. Buckingham's paper took up the one side of a debate on a recently published claim about the possibility of observationally indistinguishable universes of different sizes, a philosophically-oriented topic of the sort that has wide appeal.

Wittgenstein was interested in relativity principles by this time, for he wrote to Russell in January 1914 from Skjolden, Norway: "Doesn't the 'Principle of Sufficient Reason' (Law of Causality) simply say that space and time are relative? At present this seems to me to be quite clear; for all the events, whose occurrence this principle is supposed to exclude, could only occur at all in an absolute time and an absolute space." After presenting a thought experiment in which the principle of sufficient reason is violated, he asks Russell to think of similar cases and says "then you will see ... that no apriori insight makes such events seem impossible to us, except in the case of space and time's being relative. Please write me your opinion on this point."

So Wittgenstein was definitely wondering about some relativity principles. Although not obvious at first, building a test model of a machine on a different size scale than the intended use of the machine is basically an exercise in building a miniature situation, if not a miniature universe. As we have seen, the accompanying questions about invariance of some things under change can be formulated in terms of relativity (of the things that change). Thus, as far as the ability to appreciate points raised by thought experiments in which the sizes and properties of things are varied, subject to the constraint of effecting the same behavior, it was fortuitous that Wittgenstein had the experience of designing and building kites, spray nozzles, and propellers, as well as the benefit of discussions with Frege in 1913 about theories of symbolism. The intellectual accident of Buckingham's papers that appeared in mid-to-late 1914 showed how a theory of symbolism could be derived from understanding why models of wings and, more generally, models of the world worked when they did. The evidence that these ideas were somehow available in Wittgenstein's milieu is that they are exhibited in the (sometimes tortuous) progression of his thought as chronicled in his surviving wartime notebooks—and ultimately are reflected in the manuscript he produced at the end of the war, now known as *Tractatus Logico-Philosophicus*.

Chapter 8

A World Made of Facts

Wittgenstein had a completed manuscript with him when he was captured on the Italian Front around November 1918. In a letter to Russell, he indicates that he had completed it in August, two months before being captured. That means he completed it during a three-month leave, according to McGuinness, who estimates that Wittgenstein had been on leave from early July 1918 to the end of September 1918. He spent much of the time with his uncle Paul, who "had a genial liking for his nephew and even his nephew's philosophy"; there is a story that Paul saved him from suicide during this time. Wittgenstein's best friend from Cambridge, David Pinsent, had by now died, not in combat, but in the course of his job as a test pilot—a job with a very high accidental death rate in those early years of flying machines. Another of Wittgenstein's brothers, Konrad (Kurt), died by his own hand shortly before Wittgenstein was captured. His last remaining brother, Paul, who was a gifted pianist, had been wounded and lost an arm early in the war—a subject in many of the letters to Ludwig from his mother. At the end of the war, Ludwig was the only male in his immediate family who was not dead or suffering from permanent injuries.

After completing the manuscript that became the *Tractatus*, Wittgenstein was shortly thereafter captured, and he spent the next nine months in an Italian prison camp as a prisoner of war. He was well treated; a young man was charged with devoting himself to bringing the gaunt, weak Wittgenstein back to health. The war over, things became more relaxed in the prison camp, and he made some literate friends there, two of them schoolteachers. He appears not to have worked much more on the manuscript; he was done with it because he believed that he had solved the problems he had set out to solve. He managed to get some privileges with respect to communications and having books sent to him. He wrote to Russell: "I believe I've solved our problems finally." While still a prisoner in the camp, he managed to send one copy of his manuscript to Frege in early 1919 and another copy to Russell several months later. He was finally released and back in Vienna in late August 1919.

Wittgenstein and Russell had not seen each other since 1914, when Wittgenstein had entered the Austrian army. Meanwhile, Russell had returned to lecturing after some dramatic events at Cambridge related to his antiwar activities resulted in a prison sentence and left him without an academic home there. In early 1918, Russell delivered a series of philosophy lectures that he said were "very largely concerned with explaining certain ideas which I learnt from my friend and former pupil Ludwig Wittgenstein." In an introduction to the text of the lectures, Russell wrote: "I have had no opportunity of knowing [Wittgenstein's] views since August 1914, and I do not even know whether he is alive or dead." He mentioned Wittgenstein again at several places in the lectures. The lectures were published that same year (in the philosophy journal *Monist*) under the title "The Philosophy of Logical Atomism." These lectures of early 1918 give us a snapshot of Russell's understanding of Wittgenstein's view just prior to the crucial insight Wittgenstein had in October 1914.

In a sense, Wittgenstein's book, which would later be translated into English and given the title *Tractatus Logicus-Philosophicus*, was addressed to Frege and Russell. The book was the result of his unrelenting and torturous search for a satisfying solution to the problems that arose in his collaboration with Russell and, possibly, for satisfying answers to the objections that Frege had raised about his and Russell's proposed theory of symbolism in Wittgenstein's personal meetings with Frege. The book, he said in the preface to the work, "deals with the problems of philosophy." The phrase reflects Russell's approach to the discipline of philosophy: solving problems. The series of lectures Russell had given just before meeting Wittgenstein for the first time in 1911 was published under the title "The Problems of Philosophy." The manuscript Wittgenstein had with him when he was captured, though he wrote to Russell that he "longed" to see it published, was not written in Russell's accessible style. Wittgenstein did not attempt to make his thoughts accessible to the general reader, as Russell did. In fact, it is not entirely clear just who the intended audience of the *Tractatus* is.

Wittgenstein's purpose in writing the book was to lay out his solution to the problems of logic and philosophy. Writing was mainly about getting things right, not about getting things across. Little if anything in his writing was done for the sake of making it publishable, and, unfortunately, it may often seem to the reader, not even for the sake of making it readable. Frege's letters to Wittgenstein about the *Tractatus* have survived and have been edited and translated. Frege's first letter is full of questions about what various terms and phrases in the work mean, and he writes:

> You see, I become entangled in doubt right in the beginning concerning what you want to say, and thus just do not progress. I now often feel tired, and this hinders my understanding as well. You will not, I hope, blame me for these remarks, but rather look upon them as stimulus to make the manner of expression in your *Tractatus* easier to understand. Where so much depends on the exact comprehension of sense, one must not expect so much of the reader.

So we are in good company if we admit that we find that some passages in Wittgenstein's book are not entirely clear. Even in the original German, and to someone he admired and with whom he had discussed his ideas, as Wittgenstein had with Frege, some things in the book are not clear. Frege goes on to caution him about using different words or phrases to convey the same thing, such as "to be a fact" and "to be the case":

> On its own the use of different expressions in the same sense seems to me to be a taboo; where one does it because of a particular advantage, one should not leave the reader in doubt about it. But, where the reader, contrary to the intention of the writer, could come to connect the same sense with different expressions, the writer should point out the difference and seek to make as distinct as possible where it obtains.

In the preface, he wrote that the purpose of the book would be achieved "if it gave pleasure to one person who read and understood it." Although it was original, it was a synthesis, ideas in the synthesis coming from many places. Wittgenstein himself explicitly recognized this but was unconcerned with saying where the ideas had come from or with saying how his own ideas compared with those of others, writing in the book's introduction that "what I have written here makes no claim to novelty in detail, and the reason why I give no sources is that it is a matter of indifference to me whether the thoughts that I have had have been anticipated by someone else." He did add, though, that ". . . I am indebted to Frege's great works and to the writings of my friend Mr. Bertrand Russell for the stimulation of my thoughts."

What Wittgenstein said about what his book showed—that "the reason why these problems are posed is that the logic of our language is misunderstood"—could have been said of works by Russell and Frege as well. Wittgenstein's view was akin to Frege's, as was Russell's, in that Frege criticized the fundamental approach of subject-predicate logic, arguing that it led to all sorts of confusions and philosophical transgressions. Russell's rhetorical style in his works on philosophy of language often proceeded by first setting out a puzzle, then showing how it dissolved once the correct logical analysis of a sentence had been given. So, when Wittgenstein wrote in the *Tractatus*: "The object of philosophy is the logical clarification of thoughts. Philosophy is not a theory but an activity," he was expressing some continuity of purpose with those whose influence he did recognize.

What Wittgenstein called the sense of his book, though, had a slightly different flavor; the point of writing it was not just to convey solutions to problems in logic and philosophy of language. Put as a short summary, he said, the sense of the book was: "what can be said at all can be said clearly, and what we cannot talk about we must pass over in silence." As a general sentiment, this was not entirely novel, either. Hume had advocated consigning to the flames any book that did not meet his standards for contentfulness, and even Frege spoke of the futility of trying to decide certain kinds of questions about logic. But the expansiveness of Wittgenstein's claim and the almost-religious solemnity of its tone seem an infusion Wittgenstein brought to philosophical logic from some other part of his life.

There were certainly many important thinkers upon whom to draw besides Russell and Frege, as Wittgenstein's main biographers (Brian McGuinness, Ray Monk) and other Wittgenstein scholars (Ian Proops, Janik and Toulmin, Juliet Floyd) have discussed. The philosopher

Schopenhauer, author of *The World as Will and Representation*, was apparently very important to Wittgenstein in his youth. In their classic study *Wittgenstein's Vienna*, Janik and Toulmin note the importance that music had for him:

> . . . we have a good deal of evidence about Wittgenstein's broader interests. Music was the chief of these. Schopenhauer used to argue that the musician somehow possessed a power, which the metaphysician inevitably lacked, to transcend the limits of representations and to convey deeper feelings, attitudes and convictions, which the verbal language of formal philosophy strives in vain to express. In his more relaxed moments, Wittgenstein used in later life to discuss the expressive power of music—a topic which continued to fill him with philosophical perplexity—in terms which allotted to it little less significance than Schopenhauer had given it. Philip Radcliffe, lecturer in music at King's College, Cambridge, describes how during Wittgenstein's time as professor of philosophy at Cambridge he used to bring him bulky scores to play over on the piano [. . .] Wittgenstein placed importance on the precise manner in which the scores were played, and seemingly found in them something of the preterlinguistic significance that Schopenhauer had claimed. Nor was his range of musical tastes and interests limited or conventional. Hearing that G. E. Moore's younger son, Timothy, had organized a successful jazz "combo," Wittgenstein persuaded him to sit at the piano and explain at great length the structure and development—what Schoenberg would have called the "logic"—of jazz.

Toulmin and Janik also mention the influence of Otto Weiniger, author of *Sex and Character*, as does McGuinness. Ray Monk's subsequent life of Wittgenstein stresses Weiniger's influence in particular. Among scientists, we have already seen that Ludwig Boltzmann and Heinrich Hertz were thinkers Wittgenstein especially admired. The short summary Wittgenstein gives of the *Tractatus* hints of a state of understanding to be reached rather than an answer to be obtained. The short summary he gave about passing over in silence might well be regarded as an overarching response to, a sort of final word on, the problems that Bertrand Russell and Gottlob Frege had stimulated him to think about.

Russell's 1918 lectures on logical atomism, since they are based on discussions with Wittgenstein that took place no later than August 1914, are of some help in indicating the significance of Wittgenstein's moment of insight in September 1914, when he pondered a newspaper article about the use of models of an auto accident in a Paris courtroom. The most striking difference between Russell's lectures based on Wittgenstein's work prior to autumn 1914 and the *Tractatus* is the complete absence of any notion of a proposition as a model of reality, of a proposition as a picture, or of picturing in general in Russell's lectures. Thus, these lectures are further evidence that the intense but unarticulated insight that Wittgenstein credited with such importance really was significant in developing the presentation of his view in the *Tractatus*.

As for what Russell's lectures *do* contain, it should be understood at the outset that what the view in his lectures presents is Russell's attempt to convey those ideas of Wittgenstein's that caused him to completely overturn and revise his own views. Some of these themes, such as the theory of types, are Russell's own and are known not to be endorsed by Wittgenstein. Wittgenstein often felt he was not understood, even by Russell. With these qualifications in mind, some main points Russell makes in those 1918 lectures are nevertheless clues to the kinds of unresolved problems Wittgenstein was working on in late September 1914.

First, Russell explains that he calls his doctrine "*logical* atomism" because the sort of philosophical analysis he is propounding ends with a residual of *logical* atoms, as opposed to physical atoms. This not only emphasizes that the analysis is a logical one but also hints at an analogy with the debates about the virtues and vices of atomic models in the foundations of physics over heat and the kinetic theory of gases (statistical thermodynamics) with which both Russell and Wittgenstein were quite familiar. Another interesting point in Russell's lectures on logical atomism is the emphasis on facts. At one point, Russell reaches the conclusion that "it is with the analysis of facts that one's consideration of the problem of complexity must begin, not with the analysis of apparently complex things." He says it was only when Wittgenstein pointed it out to him that he realized the obvious: that the relationship between proposition and fact is of an entirely different sort than the relationship between name and thing named. A proposition, says Russell, is "just a symbol." And, he says, "Propositions are complex symbols, and the facts they stand for are complex." But some problems arise in figuring out how propositions are related to the facts they stand for. Again and again during the course of the lectures it turns up that, in explaining the correspondences between constituents of facts and constituents of propositions, Russell has to make special exceptions for logical constants—that is, for symbols for "or," "not," "if," and "then." He feebly resolves the issue only by assuming for the sake of argument that he has constructed a logically perfect language for general use, justifying the plausibility of being able to do so for logic in general by citing his success in doing so for the propositions of mathematics in his masterwork, *Principia Mathematica*.

Finally, there is the issue of what the world is made of, or, as philosophers sometimes put it, the question of what an inventory of the world would include. Russell says that propositions would not show up in a proper inventory of the world. What would show up are facts, beliefs, wishes, and wills. His reason for including beliefs, wishes, and wills in addition to facts seems to have to do with the problematic nature of symbolism. He says there: "I think that the notion of meaning is always more or less psychological, and that it is not possible to get a pure logical theory of meaning, nor therefore of symbolism." So, the picture we get from these lectures is that some of the more prominent problems in philosophy to be solved from Wittgenstein's standpoint, as he enters the military, are problems about the inventory of the world, the problematic nature of symbolism that any theory of symbolism has to deal with, the nature of propositions and facts, and what explains a proposition standing for a fact.

Wittgenstein felt he had solved these problems in the completed manuscript of the *Tractatus* that he was carrying with him when he was captured. In his essay "Pictures and Form," Wittgenstein's biographer Brian McGuinness remarks on the order in which some of these interrelated notions appear in the *Tractatus*. His remarks may be helpful in understanding the analogy I wish to lay out between what Buckingham's account of the methodology of experimental models says the relationship between an empirical equation and a model is, and what Wittgenstein says the relationship between a proposition as a picture and a proposition as a truth function is.

A word about the notation used in references to the *Tractatus*: paragraphs in the *Tractatus* are numbered and arranged hierarchically, with statements nested inside others. So some statements are numbered with a single numeral, such as 1, 2, 3, 4, 5, 6, or 7, and others begin with one of these numerals followed by a decimal point and further digits. Hence "the 2's" are the statements beginning with the numeral 2, "the 6's" the statements beginning with the numeral 6, and so on. Wittgenstein explained his notation as follows: "The decimal numbers assigned to the individual propositions indicate the logical importance of the propositions, the stress laid on them in my exposition. The propositions $n.1$, $n.2$, $n.3$, etc. are comments on proposition no. n; the propositions $n.m1$, $n.m2$, etc. are comments on proposition no. $n.m$; and so on." McGuinness observes:

> . . . in 6 itself the general form of a proposition is announced. By this is meant the form that any expression which is to be a proposition must have: the 4's and 5's have been largely devoted to showing that this form will be identical with the form that any truth-function must have (4's) and to showing what form it is that any truth-function must have (5's).

and later in the same essay remarks:

> Briefly the difficulties with which [the importance of "the general form of a proposition" in the sequence of Wittgenstein's thought as he has presented it to us] confronts us are the following: in the first part of the *Tractatus*, notably in the 3's and early 4's, we seem to be told that the essence of a proposition is to be a picture, while in the later parts we are told that its essence is to be a truth-function, that is to say a result of applying the operation of simultaneous negation to elementary propositions. [. . .] the two accounts seem to be quite separate things, and, if this is so, cannot both be adequate accounts of what it is to be a proposition.

McGuinness presents a puzzle in saying that the two accounts cannot both be adequate. However, this puzzle disappears once it is seen that it is in understanding how a proposition is *both* a picture *and* a truth function that the adequacy of Wittgenstein's account consists. Perhaps he would not put things exactly as I have, but McGuinness's goal is to show "the unity of Wittgenstein's account of a proposition and its form throughout the

Tractatus." By this he means the remark in 5.47 that "Where composition is to be found, argument and function are to be found also, and where these are, all the logical constants are implicit. You might say: the sole logical constant is what all propositions, by their nature, have in common with one another. This, however, is the general form of a proposition."

This is a helpful way to put things, because the significance of the general form of a proposition arises in resolving a puzzle about two things that seem very different but are "in a certain sense one." In fact, Wittgenstein introduces his account of logical form with an example showing that two things that appear to be very *different* at first have the same logical form: the musical notation of a score, and the lines on a gramophone record: "A gramophone record, the musical idea, the written notes, and the sound-waves, all stand to one another in the same internal relation of depicting that holds between language and the world" (TLP 4.014). He explains logical structure, and the relation of language and the world, in terms of what these two different languages have in common, and he does so in terms of the possibility of translation between them.

Figures 1A and 1B (which are my own sketches) show, respectively, the items and processes Wittgenstein mentions in his gramophone illustration (1A) and the facets of them upon which he draws in the *Tractatus* to illustrate the "relation of depiction that holds between language and the world" (1B).

Figure 1A illustrates the relations between a gramophone record, a live symphony performance, the musical idea the symphony evokes or expresses, the reproduction of a symphony performance from a gramophone record, the recording of a symphony performance (or a reproduction of one) in the musical notation of a score, and the production of a live symphony performance from a score. These are all things associated with producing, or producing things from, musical scores and gramophone records.

Figure 1B shows the points Wittgenstein makes in the *Tractatus* using gramophone records and musical notation to illustrate the relationship between language and the world. Since it illustrates points made in the *Tractatus*, the scheme shown is meant to reflect whatever was gained from the insight of September 1914.

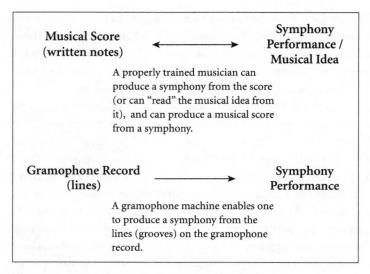

Figure 1A   Musical scores and gramophone records.

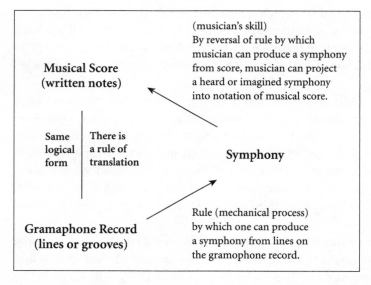

Figure 1B   Written notes and gramophone records in the *Tractatus*.

In the Tractatus, the score and the gramophone record "stand in the same internal relation" of depiction, which is what having the same logical form amounts to.

As mentioned in Chapter 1, of the four processes Wittgenstein could have drawn upon, he really only directly relies on two to illustrate what it means to have the same logical structure: the production of sound waves from the gramophone record, and the recording of this mechanically reproduced performance in the musical notation of the score. There is a subtle but important difference between the translation by Ogden and the translation by McGuinness, so I give both here:

> 4.0141   There is a general rule by means of which the musician can obtain the symphony from the score, and which makes it possible to derive the symphony from the groove on the gramophone record, and, using the first rule, to derive the score again. That is what constitutes the inner similarity between these things which seem to be constructed in such entirely different ways. And that rule is the law of projection which projects the symphony into the language of musical notation. It is the rule for translating this language into the language of gramophone records. [Pears and McGuinness translation]

> 4.0141   In the fact that there is a general rule by which the musician is able to read the symphony out of the score, and that there is a rule by which one could reconstruct the symphony from the line on a gramophone record and from this again—by means of the first rule—construct the score, herein lies the internal similarity between these things which at first sight seem to be entirely different. And the rule is the law of projection which projects the symphony into the language of the musical score. It is the rule of translation of this language into the language of the gramophone record. [Ogden translation]

The subtle but important difference between these two translations is that Ogden takes Wittgenstein to be saying that there is a *first* rule by which the musician reads the symphony from the score, and *another, possibly different*, rule by which the symphony is reconstructed from the line on the gramophone record. The German passage is not entirely clear, but I think Wittgenstein must mean that there is more than one rule here. There are two different rules by which the symphony can be reconstructed: one rule for obtaining it from the score, and another rule for deriving it from the gramophone record. Then, the rule by which the musician can construct the score from the heard symphony—which, Wittgenstein says, is "by means of" the rule by which the musician is able to "read the symphony out of the score"—means that if a musician knows how to read the symphony in the score, that ability enables the musician to project a symphony into the language of the score as well.

The ability to obtain a musical score from a symphony performance, whether live or reproduced, thus seems to be based on the *human skill* of being able to "obtain the symphony from the score." So the claim that the language of musical notation can be translated into the language of the gramophone record relies on the rule of producing the symphony from the score only insofar as it is a basis for producing the score from the symphony. This is referred to as "the law of projection which projects the symphony into the language of musical notation" (TLP 4.0141; Pears and McGuinness tr.). Or, "language of the musical score" (Ogden tr.). This makes sense if we consider that the lines of the gramophone record need encode only the features of the live symphony performance that are humanly detectable. The musical notation need not capture more than this, either. Whether the sound waves are produced from a gramophone or by a symphony performance that is produced by musicians from the score, each will have particular features that are not described by the notation, so the features that the notation pictures are appropriately thought of as only those features of a particular symphony performance that are "projected" into the musical notation. That is why the facets of the processes that display the translatability (and hence the "inner similarity" or "logical structure") between the score and the record are those facets that are associated with *both* what is reproducible from the gramophone lines and what is projectable into the musical notation of the score.

That the "written notes" that are a "picture of a piece of music"/"picture of a musical piece" (TLP 4.011; Pears and McGuinness tr./Ogden tr.) and the gramophone record, which is a very different kind of thing, are examples of things that are "in a certain sense one" sets up a general scheme of a more complicated analogy. For, the proposition as a picture and the proposition as a truth-function, are going to have the same "logical structure" or "inner similarity", too, yet they seem to be, as McGuinness points out, very different kinds of things. It is clear from the *Notes on Logic* that Wittgenstein already has some notions about propositions and symbolism in place but has not yet worked out this part of the story of how language is related to the world.

The significance of the crucial insight about models that came to Wittgenstein in September 1914 was that he began to investigate what the ideas about physical similarity that sprang up or surfaced around 1914 had to do with the problems of logic that were occupying him. (He may have already been familiar with ideas about physical similarity, albeit in the form of less-general and less-explicit accounts of the method.) What ultimately resulted from the fusion was a comparatively lean, spare account of a considered and settled view on the topics in logic that had been tormenting him. Based on intellectual content, I believe it was Buckingham's much more abstract account (rather than any of the more-specific accounts of models in any specific discipline available prior to 1914) that helped organize Wittgenstein's manuscript. From his vantage point, the way Buckingham presented the method contained an insight about what the *logic of symbolizing relations between quantities* and *the rules of correspondence between experimental models and what they model* have in common. It showed *the central role played by "the most general form of an empirical equation"* in how relations are symbolized and in how models represent. And the key thing connecting equations and models are the dimensionless parameters that characterize a system in the world. We might think of these dimensionless parameters as facts about the system, some of which we regard as fixed, and others that we regard as variable.

Referring to Figure 3, the unifying concept in the scheme is the finite collection of dimensionless parameters denoted by $(\pi_1, \pi_2, \pi_3, \ldots \pi_i)$. That collection of dimensionless parameters figures in at various places throughout the scheme. First, in the "most general form of an equation," $\varphi (\pi_1, \pi_2, \pi_3, \ldots \pi_i) = 0$, the π_is are arguments of an unspecified function. This "most general form of an equation" is not the kind of equation that scientists normally write to express a relation; rather, it is an underlying form that all (complete physical) empirical equations must have. That is, the equation as it would be stated in the mathematical symbols used in science to describe an empirical relation, such as the ideal gas law or Bernoulli's equation, would be expressible in the form $\Sigma \, M \, Q_1^{b1} \, Q_2^{b2} \, Q_3^{b3} \ldots Q_4^{bn} = 0$. These are not what Wittgenstein would call mathematical propositions; they are propositions of empirical science. In a way, we can regard this latter expression, which is a sum of products of operations (the Qs) and a dimensionless ratio (M), as a statement in a natural language— a natural language used by humans doing formal science. It is a statement of a relation that holds in the world, in the language used in science to express such relations. Thus, seeking "the most general form of an equation" is something like what Russell was doing when he sought the logical form of statements of natural language. Buckingham was doing it for the natural language used in empirical science.

Permissible and adequate sets of π_is are determined by a logical principle, the principle of dimensional homogeneity, as explained in the preceding chapter. As explained there, a dimensionless parameter π_i consists of some Qs combined in a determinate way. In other words, it is not just a list of Qs, but a ratio, such as the Reynolds Number ([density] [velocity] [distance] [kinematic viscosity]$^{-1}$), in which each Q can be thought of as interlocking with other Qs in determinate ways to form a dimensionless parameter. One way to see that they cancel out is to consider the ratio in terms of the canonical dimensions of mass, length, and time, or [M], [L], and [T], respectively. This yields [M] [L]$^{-3}$ [L] [T]$^{-1}$ [L] ([M] [L]$^{-1}$ [T]$^{-1}$)$^{-1}$, which is thus dimensionless, since all the units cancel out, or have a zero exponent. To consider a simpler example, suppose Q_1 is a velocity, Q_2 is a time, and Q_3 is a distance. Then $Q_1 Q_2 Q_3^{-1}$ forms a dimensionless parameter. The Qs interlock with other Qs only in certain ways to result in a product in which all the units cancel out, resulting in a number that has no units, or, as it is often called, a dimensionless parameter.

Second, the set of π_is is crucial in the part of the scheme indicated as "System S'" in Figure 3. System S' is "physically similar" to System S in the world. As indicated, the empirical equation mentioned describes a relation R of n different kinds of physical quantities Q. These quantities Q "pertain" to S; they are "so related by R that the value of one is fixed by the others" (think of a relation such as one of the laws relating the pressure, temperature, and volume of a gas). Often we are interested in the relation R because it involves all the quantities that determine a certain phenomenon, and it is the phenomenon that we are really interested in (think of the phenomenon of onset of turbulent flow, or of the critical point at which a substance vaporizes or liquefies). Often, the purpose of identifying or constructing the system S' is to investigate a phenomenon P in system S. The kind of practice the paper was providing an account of was one where someone builds the system S'—say, a model of an airplane wing or a screw propeller (including enough of the surrounding situation to include all the quantities related by the relation R)—and then observes its behavior. Because the purpose is actually to find out how system S will behave, the system S' is constructed to be "physically similar" to it. The system S' differs from the system S in that some of the values of the Qs are different. For instance, in building a scale model, every length might be halved; S' is "physically similar" to S if the other values of Qs are changed accordingly in such a way that all the π_is are the same in S' as they are in S. Thus, the set of π_is that occur as arguments of the unspecified function Φ in the general form of the equation are the criteria for physical similarity. Buckingham's terminology here seems to run together the system and the specification of the system, in that he talks about "transforming" the system S by altering the values of the Qs. The examples should clarify what he is saying, however, for it is quite natural to talk of "shrinking" a large object, whether actual or imagined, to a scaled-down version of it, and we may understand his talk of a "transformation" of system S this way. There is a symmetry between S and S'; there is really nothing (other than practical considerations) that determines which of S and S' should be designated the system in the world, and which a model of the system in the world. Though models are built to be physically similar to whatever they are meant to model, the mathematical relation of physical similarity is symmetrical.

Third, the π_is figure in the step where correspondences are set up between values of Qs in S' and values of Qs in S. As was emphasized in earlier chapters, one Q, such as a diameter, might correspond to a linear multiple of the value it has in S', while another Q, such as elapsed time, might vary as some power of that multiple. As explained in the preceding chapter, these correspondence rules are generally obtained by manipulation of the most general form of the equation, employing the principle of dimensional homogeneity.

Thus, the dimensionless parameters (the π_is) in the "most general form of an equation" are crucial to how the empirical equation, a system S, and a system S' are interrelated. They are used both in *describing the system by an equation* and in *modeling the system with another physical system*. The model, of course, is a model only if there are correspondence rules between it and what it models. These too are derived from the dimensionless parameters—just as a gramophone record is a gramophone record only in conjunction with a means of producing sounds from it.

The second line of Wittgenstein's *Tractatus* confidently states, without argument: "The world is the totality of facts, not of things" (TLP 1.1). The account in the *Tractatus* of facts, propositions, signs, symbols, and what the world consists of takes on a certain coherence when seen in terms of an analogy to the methodology of similarity of physical systems that Buckingham presented. In this analogy, the π_is are facts.

I will explain how these themes hang together in the *Tractatus*. The explanation is encapsulated in a set of figures, each illustrating an account of what sameness of logical structure consists in. Prior to mid-1914, Wittgenstein already had the example shown in Figure 1A (Musical Scores and Gramophone Records) available. The relationships between propositions, signs, and symbols shown in Figures 2A and 2B had already been recognized. Then the crucial insight of September 1914 occurs, in which Wittgenstein considers the possibility that a proposition works like an experimental model—whatever working like a model may be. (The crucial moment of insight is recorded in a notebook entry: "In a proposition a world is as it were put together experimentally. (As when in the law-court in Paris a motor-car accident is represented by means of dolls, etc.)") Hence, he ponders the relationship of models and empirical equations.

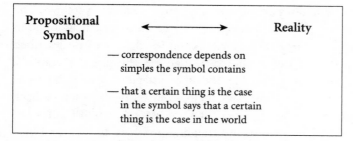

Figure 2A   Propositional symbols—prewar view.

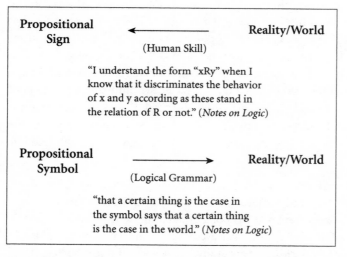

Figure 2B   Propositional signs, symbols, and truth-functions—prewar view.

Regarding propositional signs: I say that if an x stands in the relation R to a y the sign "xRy" is to be called true to the facts and otherwise false. (Notes on Logic)

Regarding propositional symbols and truth functions: "In two molecular functions that have the same T-F schema, what symbolizes must be the same." (Notes on Logic)

My speculations about what happened next are as follows. The answer to the question of how it is that models represent, as found in Buckingham's very abstract analysis and which I have just described above, is worked out as shown in Figure 3 (Empirical Equations, Model Experiments, and Physical Phenomena). Figure 3 serves as something Wittgenstein draws analogies with: he can then go back and reflect on various aspects of the processes shown in Figures 1A, 2A, and 2B. This solves the outstanding problems in logic and philosophy that had been occupying him, and he begins working out the details and consequences in his notebooks, eventually writing a manuscript in which this is all laid out. Figure 4 shows the results of working out things on analogy with Figure 3. Once he is satisfied that the problems of logic are solved, he goes on to reflect on questions about the place of logic in the world—that is, on what has been achieved in solving these problems, alternative representations in science, and the methods of philosophy, ethics, and solipsism. One of his most famous statements about the book is that its main point is ethical and consists of what he did not say in it. The occasion of the comment was Wittgenstein's inquiry about having it published. He wrote:

> My work consists of two parts: the one presented here plus all that I have *not* written. And it is precisely this second part that is the important one. My book draws limits to the sphere of the ethical from the inside as it were, and I am convinced that this is the ONLY *rigorous* way of drawing those limits. In short, [...], I have managed in my book to put everything firmly into place by being silent about it.

In the book itself, in the context of discussing what is "mystical," he says that "If a question can be framed at all, it is also *possible* to answer it" (TLP 6.5). Yet there is a sense of limitation, not of victory, in his tone, for in TLP 6.52 he writes "We feel that even when all *possible* scientific questions have been answered, the problems of life remain completely untouched. Of course there are then no questions left, and this itself is the answer." The interpretation I give here is not totally unrelated to the questions addressed later in the *Tractatus* after the questions about the nature of empirical propositions have been resolved, but the purpose of the present book is limited to explaining the proposed interpretation as it relates to the topics of propositions, logical form and structure, models, and pictures.

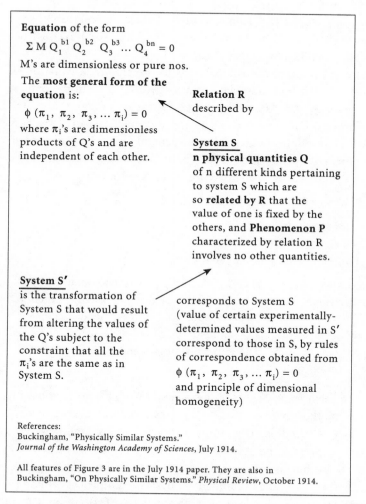

Equation of the form

$$\Sigma \, M \, Q_1^{b1} \, Q_2^{b2} \, Q_3^{b3} \dots Q_4^{bn} = 0$$

M's are dimensionless or pure nos.

The **most general form of the equation** is:

$$\phi \, (\pi_1, \, \pi_2, \, \pi_3, \dots \pi_i) = 0$$

where π_i's are dimensionless products of Q's and are independent of each other.

Relation R
described by

System S
n physical quantities Q
of n different kinds pertaining
to system S which are
so **related by R** that the
value of one is fixed by the
others, and **Phenomenon P**
characterized by relation R
involves no other quantities.

System S′
is the transformation of
System S that would result
from altering the values of
the Q's subject to the
constraint that all the
π_i's are the same as in
System S.

corresponds to System S
(value of certain experimentally-
determined values measured in S′
correspond to those in S, by rules
of correspondence obtained from
$$\phi \, (\pi_1, \, \pi_2, \, \pi_3, \dots \pi_i) = 0$$
and principle of dimensional
homogeneity)

References:
Buckingham, "Physically Similar Systems."
Journal of the Washington Academy of Sciences, July 1914.

All features of Figure 3 are in the July 1914 paper. They are also in
Buckingham, "On Physically Similar Systems." *Physical Review*, October 1914.

Figure 3   Empirical equations and experimental models—Buckingham 1914.

The rest of this book is meant to be read with the figures in mind. By the end of the chapter, or once the analogy set out in the figures is clear, the statements I make in this chapter to point out the analogy may be superfluous. It is really the analogy between Figures 3 and 4, and the ensuing understanding of how the pair of two different things that is depicted in each figure should be seen as "in a certain sense, one" (just as the gramophone record and the musical score are, in a certain sense, one) (TLP 4.014), that comprise the view I am suggesting. In fact, Figure 3 can be seen as a sort of organizing schema (perhaps as scaffolding?) used to construct Figure 4, after which it is no longer needed.

In Figure 4, the item that is analogous to the empirical equation of Figure 3 is the description of a relation in the world stated in the natural language. However, the natural language, Wittgenstein has said, often obscures a proposition's true nature, so for a clearer view of its nature, there is also the description in terms of the general form of a proposition. The general form of a proposition is $[p, \varepsilon, N(\varepsilon)]$. In his introduction to the *Tractatus*, Bertrand Russell explains Wittgenstein's notation:

> The following is the explanation of this symbol:
>
> p stands for all atomic propositions.
> ε stands for any set of propositions.
> $N(\varepsilon)$ stands for the negation of all the propositions making up ε
>
> The whole symbol $[p, \varepsilon, N(\varepsilon)]$ means whatever can be obtained by taking any selection of atomic propositions, negating them all, then taking any selection of the set of propositions now obtained, together with any of the originals—and so on indefinitely. This is, he [Wittgenstein] says, the general truth-function and also the general form of a proposition.

Thus, the form $[p, \varepsilon, N(\varepsilon)]$ is the *general form of a proposition*. My analogy between the schemes shown in Figures 3 and 4 is *analogous to the general form of an equation*. The atomic propositions are arguments of a function that expresses the general form of a proposition, just as the dimensionless products are arguments of a function that expresses the general form of the (complete physical) empirical equation. Furthermore, just as the dimensionless products (the Πs) are built from interlocking Qs in only certain ways governed by a logical principle, so (as we shall see later) are the atomic propositions built from interlocking objects that combine only in certain ways, and in an articulated manner. Furthermore, just as the Πs are independent, in that each ratio can be varied without affecting the others (they can be chosen to satisfy this requirement, as was explained earlier), so the atomic propositions ε are independent of each other, in that each can hold or not, without affecting whether any of the other εs do.

Propositional Sign

of some sort (written, sounds) it may involve signs for "and," "or." The proposition of which it is a sign also has the general form:

The general form of a proposition is:

$[\rho, \varepsilon, N (\varepsilon)]$

where ε's are elementary propositions describing relationships between Q's, and ε's are independent of each other. (nb: no logical constants occur in general form of proposition)

Humans can use language and can understand a proposition without knowing what each word stands for.

Relation R described by

World

— **n objects Q** of n different kinds are so **related by R** that the value of one is fixed by the others. **Phenomenon P** characterized by relation R involves no other quantities.

— Divides into facts ε1, ε2, ... The elementary facts ε1, ε2, ... are composed of objects in such a way that if all the ε_i's hold, then phenomenon P is unchanged.

Truth Function

The **general form of a proposition** is:

$[\rho, \varepsilon, N (\varepsilon)]$

— It is the truth function corresponding to the relation R in the world if the ε's (configurations of Q's) have the value of true if the corresponding fact in the world exists.

— It follows from the nature of a truth function that if all the truth values of ε's are unchanged, then the values of the truth function is unchanged. (Thus, the relations of the Q's affect the truth value only if the truth value of any of the ε_i's are changed, otherwise truth value is unchanged.)

corresponds to **World** (the value of the elementary proposition ε_i is "true" if fact ε_i exists)

Figure 4   Propostions in the *Tractatus.*

In Figure 4, the general form of a truth function is the same as the general form of a proposition: [p, ε, N (ε)]. As I have depicted the view in the *Tractatus*, this will be the truth function corresponding to the relation R in the world if the εs have the value of true when the corresponding fact in the world exists, and false otherwise. Thus, the correspondence rule is in terms of the εs and is simply that the value of a particular ε in the propositional truth function is "true" in the proposition if the fact ε exists (in the world). This sounds trivial, but it is key to the scheme, and its consequences are far from trivial. Correspondences at a level below the atomic propositions would be correspondences between objects. Just as the correspondences between the Qs in Figure 3 are established in terms of the fact that the Πs between the two physically similar systems S and S' have to be the same, so the correspondences between the objects, or Qs, in Figure 4 are established in terms of the fact that the atomic propositions ε that are true in the truth function exist in the world. So, just as the dimensionless parameters, the Πs, unify the scheme of how empirical equations, models, and the actual world are interrelated, so the atomic propositions, the εs, unify the scheme of how propositional signs, propositional truth functions, and the actual world are interrelated.

Another analogous aspect between Figures 3 and 4 has to do with the practical use of models and propositions. When designing model experiments, one wants to be able to manipulate the value of one of the dimensionless ratios and see how the phenomenon of interest changes. That's how we investigate modifications of the system S—by observing how the phenomenon P varies as some parameters are varied. Likewise, when studying the logical behavior of a proposition, we are interested in manipulating the values of the elementary propositions that are arguments of the truth function to see how the truth value of the propositional function varies. Thus, in both cases, there is a practical value to having a general form, in that the general form identifies the arguments that are relevant — and so permits manipulating an atomic proposition or dimensionless ratio independently of the rest, and observing the effect on the value of the propositional function or the experimental model.

The reader may see some further aspects of the analogy that have not been mentioned here; this is not discouraged. Perhaps the facets of the analogy I have tried to convey are not as clearly conveyed as I wish, in which case I can only refer the reader to the figures in hopes that more appropriate ways of drawing the analogy might occur to him or her. The point of this chapter is to convey the schemes depicted in Figures 3 and 4 and to describe the analogy between them. The points I make are valuable only inasmuch as they are helpful in seeing that analogy. Below I offer a line-by-line interpretation of the 1's and 2s of the *Tractatus*.

First, some general points. On this interpretation, objects in Figure 4 are analogous to fundamental quantities in Figure 3. How close is the analogy? Actually, a literal analogy on which the objects in Figure 4 *are* the fundamental quantities in Figure 3 is possible, but less-literal analogies are also possible. More important than the question of how literal the analogy is is the general structure exemplified in each figure that is the basis on which the analogy can be drawn. In Figure 3, which quantities are regarded as fundamental is not specified, but for empirical equations of physics, examples would be mass, length, and time, often denoted as [M], [L], and [T]. The quantities need not be restricted to quantities of physics, however. As for fundamental quantities, so for objects: although the objects are "simple" and cannot be analyzed further, this does not mean that there is a select group of them identified as simple prior to the analysis. The simplicity of objects is a matter of those objects being what one has identified at the end of an analysis, and different analyses could possibly end up with different collections of objects. In Figure 3, although there is often a canonical set in use, such as mass, length, and time, there is in principle no reason why quantities such as velocity, density, viscosity, and energy cannot instead be fundamental quantities, a point Buckingham made in the very general treatments he gave in his papers of 1914. However, an analysis in general ends up with only a few quantities as simple, the other quantities being combinations of them. So, in Figure 4, analogously, which objects are regarded as simple and which are regarded as complex is relative to the analysis. As Wittgenstein wrote in a moment of insight while working out his ideas in the *Notebooks*: "The object is simple *for me!*" I take this to mean he came to realize that he should no longer search for an objective criterion of simplicity. Instead, he came to see that simplicity is a matter of whether the analysis had come to an end (just as atomicity is a matter of not being further divisible). He also realized that many different analyses were possible, possibly leading to different collections of simple objects.

Buckingham had pointed out that, logically speaking, there was no reason to choose mass, length, and time over other possible sets of three independent quantities. And he had even shown that there are some phenomena for which only two of the quantities are independent, and some for which more than three are required.

Next come facts. The analogy here is that the facts shown in Figure 4 are analogous to the dimensionless parameters that are shown in Figure 3 having certain values. Then, in Figure 3, having those same values by certain dimensionless parameters in the system S' as in system S might be put as saying that the facts in system S' "picture" the facts in system S.

This conception of facts within the scheme shown in Figure 4 also fits well with another point made in the *Tractatus*: that objects join together in determinate ways to form facts, and that they do not need any "connectors" (logical constants) to do so. Recall that the quantities Q combine only in certain ways to form dimensionless parameters, and that dimensionless parameters are combinations of such quantities in ways so as to yield a product that has no dimension. So, in the analogy I am drawing, just as in Figure 3, the possibilities that a quantity such as mass has to combine with other quantities so as to generate a dimensionless product is determined both by it and by what the other quantities are, so in Figure 4 the possibilities that an object has to combine with other objects so as to generate a fact is determined both by it and by what the other objects are. Thus, on this interpretation of objects, we can say of objects that, though they are not "given" to us, whatever they may be, if we were to be given a set of them, the possibilities objects have for combining with each other to form facts is not relative to an analysis, but is built into the objects themselves. That is, the ways in which they combine with each other is a logical consequence of each object being what it is. It is appropriate to think of each of these objects as linking with some other objects and not others, and in determinate ways, somewhat like the notion of how atoms combine to form molecules only in certain ways that depend on their valences.

On this interpretation of objects, some of these interlocking linkages of objects form facts, and the way in which they form the fact has a structure. In other words, it is not just a pile, a list, or a mixture of the objects. Recall that the facts (εs) in Figure 4 are analogous to the dimensionless parameters (Πs) in Figure 3, and that some of the dimensionless parameters, though they are merely ratios, are ratios that have physical significance. An example is the dimensionless ratio (dimensionless parameter) mentioned earlier, the Reynolds Number. In terms of dimensions, the Reynolds Number is a product of the dimension density, the dimension velocity, the dimension length, and the dimension of the inverse of viscosity. These link together to yield a parameter that has no dimension (all the dimensions cancel out), and hence is a pure ratio. This ratio, the Reynolds Number, has physical significance in that, when the density, velocity, characteristic length, and viscosity that characterize a certain situation are used, the value of the dimensionless ratio indicates things about the situation, such as whether the flow regime is turbulent flow or a smooth, laminar flow. So, saying that the Reynolds Number is equal to a certain value, such as 10,000, is a matter of relation between the Qs, the quantities. To say that a certain atomic or elementary fact exists is to state a relation between the Qs. Then, the analogy between the εs and the Πs is tighter than an analogy; just as a dimensionless ratio having a certain value is a relation between quantities, so the existence of an ε is a relation between quantities as well.

In experimental modeling, characterizing a situation often requires more than one dimensionless ratio. It can be proven that a set of dimensionless ratios that characterize a situation can always be chosen such that each dimensionless ratio is independent of the others in the set—that is, such that each dimensionless ratio can be varied while the others in the set stay the same. My interpretation of facts as analogous to dimensionless ratios fits in this respect, too. For, facts are configurations, and the existence of an exhaustive collection of the facts characterizes situations, just as collections of dimensionless ratios are used to characterize a physical situation, and provide criteria by which physical situations that are similar to it can be identified.

On this interpretation of objects, atomic (or elementary) facts, and facts, the statements in the *Tractatus* about these terms and about propositions, models, and picturing come out as quite natural things to say. I leave it open how literally or analogously the interpretation should be taken, but note that dimensional analysis is a very general method, as applicable to biology as to physics, and even applicable to architecture, economics, and sociology, so a closely analogous interpretation is not implausible. Next I show how natural many of the statements in the *Tractatus* come out on the interpretation I have just proposed.

§§§§

1   Die Welt ist alles, was der Fall ist.

The world is everything that is the case. (Ogden tr.)

The world is all that is the case. (Pears and McGuinness tr.)

1.1   Die Welt ist die Gesamtheit der Tatsachen, nicht der Dingen.

The world is the totality of facts, not of things. (Ogden tr.)

The world is the totality of facts, not of things. (Pears and McGuinness tr.)

The facts that exist (on analogy to the dimensionless parameters' having certain values) characterize a situation. Thus, just as the values of dimensionless parameters (Πs) divide the world into equivalence classes (relative to a phenomenon of interest) characterized by the values of those Πs, so the existence of facts divides the world into equivalence classes of situations that are characterized by the existence of those facts. The idea is not unlike the one that arises in statistical thermodynamics, in which there are many different configurations of molecules that will yield the same macroscopic thermodynamical phenomenon, and the configurations that yield same phenomena are all put into a single equivalence class on that criterion. What is relevant for empirical science is whatever it is that all the configurations in that equivalence class have in common. Later in the *Tractatus*, in statement 5.5423, Wittgenstein writes that "To perceive a complex means to perceive that its constituents are related to one another in such and such a way" (Pears and McGuinness tr.) or "To perceive a complex means to perceive that its constituents are combined in such and such a way" (Ogden tr.). Then, referring to a Necker cube, he remarks: "This no doubt explains why there are two possible ways of seeing the figure as a cube; and all similar phenomena. For we really see two different facts" (Pears and McGuinness tr.).

This view contrasts with one on which the world consists of things (such as molecules or material points) or even of the measurements one could take of things in it. Instead, the world consists of facts, these facts being relations or ratios rather than things or values of quantities. Ratio is a good concept to have in mind here, since the ancient and more general notion of the concept includes relations of a nonmathematical sort. So, for instance, analogies such as state:citizen :: parent:child are akin to statements of numerical ratios such as a:b :: c:d, which we now write in modern arithmetical notation as $a/b = c/d$. In Figure 3, examples of a dimensionless parameter might be inertial force : viscous force, and an example of a fact would be equating this ratio to a number. Saying that the world is the totality of the values of those ratios, as opposed to an inventory of things, is central to the view being put forth, and is a natural thing to say on my interpretation, since it is such ratios, rather than particular things or values of particular quantities, that determine phenomena.

The notion that the world consists of ratios rather than things, inspired by an understanding that the phenomena of science are explained by ratios (and that varying ratios correspond to varying phenomena), has a well-known precursor: Pythagoras of ancient Greece. Again, the key example is an example of alternative representations of musical performances. The canonical example associated with Pythagoras was that of playing musical tones on a single-stringed instrument. Legend has it that Pythagoras was inspired to develop a view that the world was composed of numbers by observing that there is a relationship between the ratio of the length of a string permitted to vibrate and the musical interval between the notes produced by the vibrating string. Helmholtz, a philosopher-scientist of Wittgenstein's era, developed the mathematical basis of tone further, in terms of frequencies of sound waves, but he too explained musical phenomena in terms of ratios (a notion called consonances). Using a broader notion of ratio would broaden the Pythagorean claim to something like the claim that the world consists of relations, not things. We shall see that the existence of facts is the existence of relations between objects, so the interpretation here is rather like a Pythagorean view with a broadened notion of ratio.

1.11 Die Welt ist durch die Tatsachen bestimmt und dadurch, dass es alle Tatsachen sind.

The world is determined by the facts, and by these being all the facts. (Ogden tr.)

The world is determined by the facts, and by their being all the facts. (Pears and McGuinness tr.)

The specific ratio called the Reynolds Number was described earlier; dozens, perhaps even hundreds, of such dimensionless ratios are important enough to be given a name. They are used not only in hydrodynamics, but in other scientific and engineering fields as well, and their values are often used to determine which side of a critical transition a system is on. Just as the value of a Reynolds Number determines the qualitative feature (phenomenon) of smooth or turbulent flow in a system, the value of a Richardson Number determines whether an underground stream entering a lake will mix with the lake water or continue as a coherent stream. To generalize this would be to say that the phenomenal features of the world are determined once all the facts (relations between objects) that characterize it are determined. This reflects Buckingham's remarks about a model of the world; that is, his point that, as long as all the dimensionless parameters relevant to the phenomena we experience are the same, we would get a world like this one, as far as our experience goes. This is a requirement of exact or total similarity. It is very seldom achieved in practical experimental modeling, but, logically speaking, the phenomena of the world are determined once one has specified a set of dimensionless parameters and determined their values. The analogous claim for Figure 4 is that, if all the facts were the same, no matter what the values of the objects, the world would be the same so far as we could tell. Hence, the world is determined by the existence of all the facts.

The stipulation "and by these being all the facts" likewise is just what one would say in this interpretation. Knowing how to characterize a situation includes knowing that the set of ratios you have used to characterize it is all that is needed to characterize it—that there are no other unspecified degrees of freedom, so to speak. You need a complete set of generalized variables to determine the behavior of the system, and you need to know that it is complete.

1.12 Denn, die Gesamheit der Tatsachen bestimmt, was der Fall ist und auch, was alles nicht der Fall ist.

For the totality of facts determines both what is the case, and also all that is not the case. (Ogden tr.)

For the totality of facts determines what is the case, and also whatever is not the case. (Pears and McGuinness tr.)

This statement elaborates on the previous ones. Since "the world" is "all that is the case," this statement is a more inclusive claim about what is determined by the facts than statement 1.11 is. If something is the case, it will show up somewhere among the totality of existent facts. 1.11 is a sort of statement of completeness of the collection of facts ($\varepsilon 1, \varepsilon 2, \ldots \varepsilon \iota$), in analogy (refer to Figure 3) to the sufficiency of the collection of dimensionless parameters ($\pi_1, \pi_2, \pi_3, \ldots \pi_i$) to completely characterize the phenomenon of interest. (Recall that the π_is (which are analogous to facts) result from forming a set of independent dimensionless parameters from the n physical quantities Q of n different kinds.)

1.13 Die Tatsachen im logischen Raum sind die Welt.

The facts in logical space are the world. (Ogden tr.; Pears and McGuinness tr.)

The puzzling phrase "facts in logical space" has a natural explanation in this interpretation. Buckingham's paper emphasized that dimensionless quantities may be expressed in terms of basic dimensions in a variety of ways, depending on which dimensions are chosen as basic. Thinking of facts as in logical space is a way to think of facts without regard to the specific choice of fundamental quantities (just as we can refer to a parabola as a curve of a certain shape rather than expressed in terms of a particular choice of coordinate system and point of origin). Thinking about the facts without regard for the choice of fundamental quantities is thinking about them without regard for what the objects of the world are. The notion is common enough in mathematics: we speak of invariants (such as determinants of a matrix) that are the same in every choice of coordinate system. We saw earlier that some physicists focused on which quantities were basic or fundamental, whereas Buckingham's paper, taking a much more abstract and logical view of the matter, treated the fundamental or basic dimensions as involving some choice. Hence, it makes sense to speak of facts in logical space rather than in terms of the kinds of objects they may consist of, which would involve arbitrary choices that are irrelevant to the fact being the fact it is. It is not uncommon today to refer to a set of dimensionless parameters as a "configuration"—which evokes the practice in physics of defining a configuration in configuration space—and conflating talk of it with talk of the physical situation.

Even so, the statement here in 1.13 is a rather radical statement about the world, even on this interpretation. It goes way beyond saying that the world, or what is the case, is determined by the facts; it says the facts in logical space *are* the world. The way I make sense of this statement is that it hints at the type of analysis of the world that the book is going to be discussing. The statement then could be seen as making the point Russell made in his exposition of a view that he said was largely due to things he had learned from Wittgenstein: that the type of analysis he was propounding was a logical analysis, and that the result of the analysis would be *logical* atoms as opposed to *physical* atoms.

1.2   Die Welt zerfällt in Tatsachen.

The world divides into facts. (Ogden tr.; Pears and McGuinness tr.)

That the world splits into facts (not necessarily atomic ones) is just what one would say in the interpretation resulting from the analogy I am suggesting. For, a system is characterized by a set of dimensionless parameters. (In other words, the state of a particular system is characterized by the values held by those dimensionless parameters, some of which can be varied, and others of which are fixed and invariable throughout.) On my interpretation, the facts that exist and hence characterize the world are (or are analogous to) the dimensionless parameters associated with the general form of any complete physical empirical equation taking on certain values. Choosing an independent set of dimensionless parameters (of the fewest number required) is closely analogous to characterizing the world in terms of the fewest number of independent facts required to achieve that.

1.21 Eines kann der Fall sein oder nicht der Fall sein und alles übrige gleich bleiben.

> Any one can either be the case or not be the case, and everything else remain the same. (Ogden tr.)

> Each item can be the case or not the case while everything else remains the same. (Pears and McGuinness tr.)

The theorem in Buckingham's paper addressed the question of how many dimensionless parameters were required to characterize a system, and so involved finding a logically independent set of dimensionless parameters of minimal size.

The criterion of independence of dimensionless parameters ensures that, in Figure 3, the value of each independent dimensionless parameter in the set can be changed without affecting the others in the set. If whether something is the case or not—such as whether there is turbulence or not, whether there is a sonic boom or not, whether a certain gas is liquefied or not—depends on the value of one of these dimensionless parameters, then it is entirely natural to say that, accordingly, each such phenomenon can exist or not exist, while all the others stay the same. This statement, 1.21, just says the same for Figure 4, for the facts into which the world divides. For facts, variation can only mean that a fact that existed or obtained no longer does (that something that was the case is now not the case) or that a fact that did not previously exist comes to exist (that something that was not the case now is the case). Thus, 1.21 is part of establishing the analogy between Figures 3 and 4.

2   Was der Fall ist, die Tatsache, ist das Bestehen von Sacherverhalten.

> What is the case, the fact, is the existence of atomic facts. (Ogden tr.)

> What is the case—a fact—is the existence of states of affairs. (Pears and McGuinness tr.)

"*Tatsache*" means "fact," and the added association here with "what is the case" indicates that the fact exists. On my interpretation, this simply reflects that the existence of a fact is analogous to a dimensionless parameter's having a certain value. A more literal interpretation than an analogy is also possible here—an interpretation on which a fact's existence *is* a dimensionless parameter's having a certain value. Or, an interpretation on which, for some facts, the fact's existence just is a dimensionless parameter's having a certain value, but that other kinds of facts are possible as well.

This is the first time that the term "*Sachverhalten*," translated as "atomic facts" by one translator and "states of affairs" by another, appears in the *Tractatus*. *Sachverhalten* has connotations of being compact and elementary in the sense of not being further decomposable into smaller items of the same kind, and basic in the sense of other things being explained in terms of it. On my interpretation, an atomic fact (or "state of affairs") is a dimensionless parameter that occurs in the collection used to characterize a system having a certain value. As we saw earlier, the values of dimensionless parameters are often used to indicate whether a critical point has been reached, such as the onset of a turbulent flow regime or liquefaction of a gas. Once a collection of independent dimensionless parameters of the minimum number required to characterize a given system is chosen, the phenomenon P is determined.

However, there is another possible distinction that Wittgenstein might have had in mind in distinguishing between facts and elementary facts. It is particularly interesting, because it would explain a discomfort he later came to feel about claims he had made about atomic or elementary propositions in the work. In Figure 3, the Πs can be chosen such that they are independent; this is established in Buckingham's paper as a matter of logic. However, although Buckingham did not do so in his 1914 papers, it is possible to draw a distinction among Πs: there is a sense in which some dimensionless ratios are compound and some are elementary. Most ratios can be thought of as eigenratios—ratios of two quantities. For example, most ratios in hydrodynamics are ratios of force, such as the ratio of inertial force to viscous force. Many of these can be decomposed into products of ratios of the basic quantities, but not all of them can. It would be understandable to make the mistake of thinking that all dimensionless ratios can be decomposed into eigenratios of the quantities chosen as fundamental ones, and that each of these eigenratios is independent of the others. Since

Buckingham's abstract formal approach had only recently been presented when Wittgenstein wrote the *Tractatus*, and Wittgenstein's research experience in Manchester would have been with much more specific and practical forms of the method of experimental scale models, this subtle point might well have gone uninvestigated. One famous example that might lead one to think decomposition into elementary eigenratios was unproblematic was Onnes's formulation of the law of corresponding states, which he discussed in his December 1913 Nobel lecture. Onnes's approach was to define eigenratios—the ratio of temperature to critical temperature, the ratio of pressure to critical pressure, and the ratio of volume to critical volume. So it might have seemed natural that this was true in general. In fact, strictly speaking, it is *not*; while the logical principle of dimensional homogeneity does guarantee that a set of independent Πs can be obtained, it does not guarantee that the Πs can be decomposed into a set of independent eigenratios of the basic quantities (which are analogous to ratios or relations of objects of the same kind). What is especially interesting about this suggestion is that a number of years after the *Tractatus* was published, it was his views in the *Tractatus* about elementary propositions that led Wittgenstein to think that some of his views in the *Tractatus* were wrong. It is to his credit that the major claims he makes in the 1s are about facts rather than about atomic or elementary facts. On the interpretation of elementary facts as eigenratios of fundamental quantities, the 1s would not necessarily hold were they made about atomic or elementary facts rather than about facts.

Another suggestion in the literature is that the *Sachverhalten* are like the configurations or states of affairs in statistical thermodynamics. What is interesting is that the solution Wittgenstein comes up with does not side with one or the other of these interpretations, for facts can relate equations and models, in the ways depicted on a literal analogy between Figures 3 and 4. So the interpretation I am proposing does not exclude an interpretation on which the facts characterize such configurations. The interpretation of facts as dimensionless parameters or ratios in a broad sense is very comprehensive. The law of corresponding states showed how the collection of the three dimensionless ratios (reduced temperature, reduced pressure, and reduced volume) characterizes the state of a substance and how Onnes regarded these states in terms of equivalence classes of physically similar dynamic models of the kinetic theory of gases.

Given that this interpretation is based on Buckingham's paper, and Buckingham's special expertise was thermodynamics, it is not surprising that what this interpretation makes of Wittgenstein's terminology of facts and objects accommodates concepts of statistical thermodynamics so easily. One might not at first expect it of someone who had been a student of Ostwald, as Buckingham was, who is known for having pushed the "energetics" approach against Boltzmann's models. On the other hand, Henk Visser has pointed out that Ostwald and Boltzmann seem to have affected each other. He even observes that "Ostwald had anticipated Boltzmann when he wrote that 'all scientific representations rest on the construction of assigned symbols of the same variety or multiplicity as that which was represented and the lawful conservation of those symbols in the place of real things.'"

The 2s that fall under this statement ("2 What is the case, the fact, is the existence of atomic facts [states of affairs]") are remarks on facts and objects that should seem natural on the proposed interpretation by now, if by atomic facts or state of affairs we just mean dimensionless parameter or elementary proposition. For example, 2.01 is the statement that atomic facts or states of affairs are combinations of objects: in Figure 3, the Πs are combinations of quantities; in Figure 4, the εs are relations of objects, as we saw earlier. The points in the remarks under 2.01, such as that it is essential to things or objects that they can be constituents of atomic facts or states of affairs, that "If I know an object I also know all its possible occurrences in states of affairs," that "If all objects are given, then at the same time all *possible* states of affairs are given," and so on have already been discussed as well.

One remark under the 2.01s bears special mention: 2.0141 "The possibility of its occurrence in atomic facts is the form of the object" (Ogden tr.); "The possibility of its occurring in states of affairs is the form of an object" (Pears and McGuinness tr.). This is the flip side of the claim that it is essential to objects that they can be constituents of atomic facts or states of affairs. On the interpretation of objects as quantities such as velocity or viscosity, 2.0141 says that the form of the object is a matter of how that quantity can combine with others to form facts. It is a natural thing to say, for, suppose that all we knew about an object was how it combined with all of the other objects. (This is not a practical situation, for objects are not given to us; they are the result of an analysis.) As an exercise in logic, what if we were told how one of the objects could combine with each of the others, but not what it was? Actually, using dimensional analysis, we could in fact determine what the dimensions of the object was. That, then, would be what its form was. So, the form of an object might be length, though we might not be able to say whether we are talking about the diameter of a pore in the wall of a container or the mean free path of a molecule. This is intriguing, for it distinguishes the form of an object from the object. In the methodology of scale models, it is a subtle distinction that one would only make when reflecting on the logic of the method. It may well be a point that Wittgenstein makes that clarifies Buckingham's treatment.

2.02 is "Objects are simple" (Pears and McGuinness tr.) and "The object is simple" (Ogden tr.). As discussed previously, on the proposed interpretation, this is not to be taken as simplicity in any absolute sense, but is understood as a matter of objects being the result of a completed analysis. Statement 2.0201, "Every statement about complexes can be resolved into a statement about their constituents and into the propositions that describe the complexes completely," indicates this is Wittgenstein's point here as well. In Figure 3, objects are quantities, and they can be understood on analogy to how quantities are used in Buckingham's paper: they are both the Qs in the Πs that appear in the most general form of an equation and the physical quantities in the world. In Figure 4, objects are the Qs that occur in the εs that appear in the most general form of a proposition, and also the objects in the world. The 2.02s involve a difficult concept that has not been discussed so far, however: the notion of substance.

2.021   **Die Gegenstande bilden die Substanz der Welt. Darum konnen sie nicht zusammengesetzt seing.**

Objects form the substance of the world. Therefore they cannot be compound. (Ogden tr.)

Objects make up the substance of the world. That is why they cannot be composite. (Pears and McGuinness tr.)

2.0211   **If the world had no substance, then whether a proposition had sense would depend on whether another proposition was true. (both translations)**

It is significant that Wittgenstein never says what objects are, though he says a great deal about them. The one time he does seem to give an example of an object, in 4.123, is of a pair of objects—two shades of blue. This is, of course, a famous example of Hume's, and for Hume, shades of blue were impressions. It is notable that Wittgenstein in that passage adds a parenthetical remark about "the shifting use of the word 'object.'" The parenthetical remark should warn us not to take shades of blue as quite the kind of things that objects are on his account. It is notable that in this one example he gives, objects are values of quantities—shades of blue. He never talks about objects being points, either as locations in space or as Hertz's "material points."

It is also important to keep in mind that nothing in his account requires that the objects that result from an analysis are unique. In fact, his later remarks about the different "nets" that can be used in Newtonian mechanics to describe the world are more in keeping with a realization on his part that in fact the fundamental quantities of a science are not unique, that there can be some room for choice.

The first remark quoted previously, statement 2.021, that objects cannot be composite, just reflects that objects are what one is left with at the end of an analysis—at the end of a process of breaking down composites into the things of which they are composed. So it reflects that the analysis can go no further. The second statement, 2.0211, is related; it reflects that the analysis went as far as it did. The idea is that, were composition into objects not possible, we would not have a level lower than the level of the elementary proposition. The proposition has sense in virtue of being composed of objects; what if no such level of analysis could be performed? The answer is that, then, the proposition could have sense only in virtue of its relations to other propositions. In this thought experiment, then, if all the other propositions to which it were related (by relations of consequence) were false, it would have no sense. What an odd kind of contingency! It's a reductio ad absurdum; the world does have substance. That's the point: whether a proposition has sense shouldn't be contingent. Hence the significance of the analysis being able to go as far as it does—all the way to objects. If whether a proposition has sense is contingent in this way, how would we be able to describe the world via propositions? Wittgenstein says, in 2.0212, that we couldn't even sketch a false picture of the world. Yet we do sketch pictures of the world, even false ones. False pictures of the world are imagined worlds. What both the imagined world and the actual one have in common are these objects, as he writes in 2.022. Next is statement 2.0231:

2.0231   **Die Substanz der Welt kann nur eine Form und keine materiellen Eigenschaften bestimmen. Denn diese werden erst durch die Satze dargestellt—erst durch die Konfiguration der Gegenstande gebildet.**

The substance of the world can only determine a form and not any material properties. For these are first presented by the propositions—first formed by the configuration of the objects. (Ogden tr.)

The substance of the world can only determine a form, and not any material properties. For it is only by means of propositions that material properties are represented—only by the configuration of objects that they are produced. (Pears and McGuinness tr.)

On the interpretation indicated in Figure 4 on analogy to Figure 3, phenomena P are determined first by the relations described by propositions, not at the level below. The canonical example here for Figure 3 would be that the phenomenon of turbulence is not something that can be characterized by the value of any quantity. For instance, early attempts to characterize turbulence as a function of velocity failed. Turbulent flow is, however, characterized by the value of the Reynolds Number— that is, by a dimensionless parameter. So, on this interpretation, the analogous point in Figure 4 would be just what the second sentence of 2.0231 states.

2.024   **Die Substanz ist das, was unabhangig von dem was der Fall ist, besteht.**

Substance is what exists independently of what is the case.

2.025   **Sie ist Form und Inhalt.**

It is form and content. (Ogden tr.; Pears and McGuinness tr.)

These statements fit well with the proposed interpretation. Objects, which we have just been told form the substance of the world, are quantities. Which ones are chosen as basic is the result of an analysis and does not depend on contingencies such as what is the case. Referring to Figure 4, this means that objects do not depend on the existence of facts for their existence. This is consistent with the previous remark that what imagined worlds and the actual world have in common is their form, the objects.

2.026   Nur wenn es Gegenstande gibt, kann es eine feste Form der Welt geben.

Only if there are objects can there be a fixed form of the world. (Ogden tr.)

There must be objects, if the world is to have unalterable form. (Pears and McGuinness tr.)

In keeping with the remarks that precede it, if the analysis could not be carried to the level of decomposing elementary facts into relations between objects, we would be stuck at the level of elementary facts, and these can exist or not, contingent upon how things are in the world. Whether the quantity has some value or other is not contingent (as Wittgenstein emphasizes in 2.0131: "Notes must have *some* pitch, objects of the sense of touch *some* degree of hardness, and so on."). So, the quantities are the unchangeable underlying format. The rest of the 2.02s elaborate on this point.

The 2.03s are about objects and states of affairs or atomic facts. I have already discussed this aspect of the proposed interpretation, but I want to mention two points about the 2.03s. Here is where Wittgenstein explicitly says that objects fit into one another like links in a chain in the atomic fact or state of affairs, and that form is the possibility of structure. The metaphor between links in a chain and the quantities of a dimensionless ratio is apt. What was especially important to Wittgenstein was that no connector is required; the objects just fit into each other, just as the links of a chain don't require glue or connectors to link them together. Given that logical constants such as "and" were something that Russell hoped to eliminate, and that Wittgenstein regarded as symptomatic that a theory of symbolism was incorrect, this is crucially important to both of them. Secondly, as mentioned in earlier chapters, on the proposed interpretation, the form of the objects allows for many different kinds of combinations. Hence, it is quite natural to say that the objects, which determine form, give rise to the structure of these facts or dimensionless parameters. Or one might well put it aphoristically that form is the possibility of structure. Statements 2.04 and 2.05 echo points made earlier about facts, but specifically for atomic facts or states of affairs. In 2.06, the notion of reality is brought in and explained in terms of the existence and nonexistence of states of affairs. In 2.063, the notion of the world is given in terms of reality: "The sum total of reality is the world."

The 2.1s are about pictures and picturing. Now all the points I have laid out come into play to explain what pictures are and how they picture what they picture. Wittgenstein says lots and lots of things about pictures in the 2.1s. Pictures are facts, but also that we "picture facts" or "make to ourselves pictures of facts." Also, "A picture is a model of reality" and it "reaches up" to reality. A picture is not only a fact, but it includes the "representing relation" or "pictorial relationship" that makes it a picture.

2.1   Wir machen uns Bilder der Tatasachen.

We picture facts to ourselves. (Pears and McGuinness tr.)

We make to ourselves pictures of facts. (Ogden tr.)

The statement involving "ourselves" occurs only after "reality" has been introduced into the conversation. Picturing is an activity, and by saying that we picture things *to* ourselves, Wittgenstein is saying not only that we can and do make the pictures of facts, but that we are able to "read" (understand) the picture too. In Figures 1B, 3, and 4, pictures of facts are shown: the musical score in Figure 1B, the empirical equation in the natural language used by scientists in Figure 3, and the propositional sign in Figure 4. What is picturing? As one scholar recently put it, and as I have illustrated in Figure 4, based on a remark Wittgenstein makes in the 4s, it seems that for Wittgenstein being a picture is characterized by the feature of being able to understand the sense of a propositional sign without knowing what each word stands for. In the subsidiary statement 2.11, he says that "A picture presents a situation in logical space, the existence and non-existence of states of affairs." This need not mean that the picture itself asserts the existence or nonexistence of states of affairs, but that it depicts the existence of nonexistence of states of affairs. There is the further question of whether what the picture depicts is actually how things are.

2.12   Das Bild ist ein Modell der Wirklichkeit.

The picture is a model of reality. (Ogden tr.)

A picture is a model of reality. (Pears and McGuinness tr.)

What does "is" mean in "A picture is a model of reality"? Referring to Figures 3 and 4, on my analogy, the truth function is analogous to a model, and the propositional sign is analogous to an equation. The analogy was set up this way because a truth function is like a model in that the arguments of the function can be varied independently and the effect on the value of the truth function calculated. The arguments of the truth function corresponded to things in the world, on analogy to a model, as illustrated by comparing Figures 3 and 4 in this respect.

Here Wittgenstein is saying that a *picture* is a model of reality, though. What is he up to here? The model-like aspects of a picture he picks out in the subsequent statements in the 2.13s and 2.14s are that the elements of the picture correspond to objects. On the interpretation presented in Figure 4, objects Q are related by a relation R so that the value of one Q is determined by all the others. The relation R is described by the propositional sign. This seems to say that the relation R that relates the objects Q is represented in the propositional sign. The example to have in mind here is musical notation. The signs used in musical notation stand in certain relations, and that is what its pictorial form consists in. This Wittgenstein calls the structure of the picture.

There is structure, and there is the possibility of structure. The structure of the picture consists in the relations between its elements. The possibility of the structure is form, or the form of representation. We don't normally think of pictures as having a form of representation, but when we do, we think of them as being structured, such as by the use of symbols or symbolism. By saying that a picture is a model of reality, Wittgenstein is saying that the pictures we can understand without having their sense explained to us have the kind of structure that a model does. We saw that, on my interpretation, what corresponded in a model to whatever it was it modeled were ratios, or relations. This fits extremely well with the remarks about picturing here, and explain that what Wittgenstein means in saying that a picture is a model is that even though we can understand the picture without having its sense explained to us, what accounts for its ability to picture is the very same thing that accounts for a model's ability to model: corresponding ratios, or relations. This is exactly what he says in 2.15: "The fact that the elements of a picture are related to one another in a determinate way represents that things are related to one another in the same way" (Pears and McGuinness tr.). The possibility of structure is the form of representation. It is this, and not the sameness of structure, that he refers to when he says "*That* is how a picture is attached to reality; it reaches right out to it" (TLP 2.1511; Pears and McGuinness tr.). So it is not that the whole account of picturing is to identify a structure in the picture with the same structure in the world and say, "It matches!" Rather, Wittgenstein's account is that we match up a form of representation (which allows for various structures to obtain), which is what makes it possible for something to be a picture of what it pictures; then there is the further question of whether or not the picture "matches" reality. As he says in 2.17, "What a picture must have in common with reality, in order to be able to depict it—correctly or incorrectly—in the way it does, is its pictorial form." A picture by itself can't be true or false (as he says in TLP 2.224, "It is impossible to tell from the picture alone whether it is true or false."). We must compare it to reality, and that requires some rules about how the elements of the picture correspond to the things in reality. Here the analogy with physical models is very helpful, and the answer for physical models is exactly the answer he gives: there are correlations between the elements and the objects.

All we need to ask is, if it is the sameness of ratios that make the picture a true one, what is it that establishes what the objects stand for? If the picture is to work like a model that is a model in virtue of physical similarity, then, referring to Figure 3, the rules for correspondence of quantities Q (which are analogous to the objects in Figure 4) and quantities in the model can be derived from the relevant ratios that are the same in the model as in the world. I will not go into detail here, except to say that it is a matter of fairly simple algebra, and that in experimental engineering models, they are called correspondence rules. These are not scientific laws of any sort; they are just simple arithmetic rules used to get from the observed quantities in the model to the corresponding values in the thing modeled. Wittgenstein's points about pictures appeal to these model-like features of pictures. For one, a model does not inherently model anything; it is just a physical thing or system. What makes it a model is the recognition of the ratios that it has in common with something else, and these ratios also yield the correspondence rules that permit translating the values of quantities in the model to what it models. That's what "understanding" the model consists in, and that's what makes it a model.

So, just as we need to think of the physical model *along with* these correspondence rules (which are derived from the ratios that are the same in the model as in reality), so for a picture. Statements 2.1513, 2.1514, and 2.1515 say "So a picture, conceived in this way also includes the pictorial relationship, which makes it into a picture. The pictorial relationship consists of the correlations of the picture's elements with things. These correlations are, as it were, the feelers of the picture's elements, with which the picture touches reality." In correspondence with the translator, C. K. Ogden, Wittgenstein said he wanted to indicate something like a butterfly's antennae, and Ogden chose the word "feelers." I think this means there needn't be the kind of similarity that photographs or sketches have; rather, only the kind of similarity that permits making correlations of selected elements of the picture with things in reality is required.

These model-like aspects of a picture explain something that is otherwise hard to explain: Wittgenstein says both that "What constitutes a picture is that its elements are related to one another in a determinate way" (TLP 2.14) and *also* "The pictorial relationship consists of the correlations of the picture's elements with things" (TLP 2.1514). These seem to provide two different accounts of what being a picture consists in. The analogy sketched in Figure 3 shows that, yes, these are two different claims, but both the physical similarity and the correspondence rules are a matter of sameness of the ratios, so the two accounts are really two facets of one requirement.

Pictures are models (or model-like) for the reasons just explained. But are pictures and models the same thing? It does not follow from the statement in the *Tractatus* "A picture is a model of reality" (TLP 2.12) that all models are pictures. All this says is that a picture works like a model in that the relations between the elements of the picture represent the relations between things in reality. But if, as the Wittgenstein scholar Ian Proops reads the *Tractatus*, "the ground for saying that the proposition is a picture [. . .] is that we can understand the sense of a propositional sign without having had it explained to us," it seems that it is a picture-like feature of the proposition that, as Wittgenstein puts it in 4.021, "A proposition is a picture of reality: for if I understand a proposition, I know the situation that it represents. And I understand the proposition without having had its sense explained to me."

We can now easily make sense of Wittgenstein's remark that a proposition is like a yardstick that it is laid against reality like a measure (2.1512) For, when the correlation is made from an element of reality to an element in the picture or model, the picture or model is used as a device to measure how things are in the world. In fact, the same can be said of physically similar systems: a system can serve as a measuring instrument of a system to which it is physically similar. Think of how a color chart might be used: when the shade on the wall is found to match with the shade of one of the color samples, then we have used the color chart to "measure" the color of the wall. So, too, whenever a picture or model is used to indicate how things are in the world: when the model behaves like the thing it is supposed to model, then it is physically similar and can be used to say how things stand in the world. Because a proposition is used to say "This is how things stand," it is natural to say that it is being used as a measure.

Then come the more philosophical points about what cannot be depicted on this account. Wittgenstein writes that a picture cannot depict its pictorial form (2.171). This is certainly true of physical models as well. The only way one can depict that the ratios in the model are the same as in what it depicts is by comparing the model and what it models, and that is not something that can be done within the model. Likewise for a picture: "A picture cannot . . . place itself outside it representational form" (2.174). Wittgenstein's reasoning is that representation requires placement outside what is represented, so pictures and models can't represent their own form. This is a prevention of circularity in representation inherent to his account; this kind of circularity, we have seen, was the problem that Russell meant to address with the theory of types. Here the theory of types is not needed to prevent the circularity of something representing itself— just what Wittgenstein had hoped for in his account. Later, after discussing the propositional sign in the 3s, he writes "No proposition can make a statement about itself, because a propositional sign cannot be contained in itself (that is the whole of the 'theory of types')" (TLP 3.332).

The preceding commentary on the 1s and 2s is meant to describe the interpretation I propose, which I speculate was inspired by thinking about how experimental engineering models represent what they do. The description is a matter of conveying the analogy I see between Buckingham's account of physical similarity in Figure 3 and Wittgenstein's account of propositions in the *Tractatus*. Conveying analogies in words is always less than satisfying, and there is the risk of a sentence's being taken to make a claim rather than to point out an aspect of an analogy. I hope by now I have said enough to convey the analogy I wish to draw. Referring again to Figures 3 and 4, it is now understandable why, in light of the account of physical similarity on which (i) "the most general form of an equation" is the unifying concept between equation and model, (ii) having the same ratios or relations is what a model and what it models have in common, and (iii) correspondence rules can be derived from those ratios, physical similarity was something Wittgenstein would latch on to and try and squeeze something out of in hopes of solving the problems he was thinking about in 1914.

The drama in all of this is that Buckingham's treatment, which, for a paper on engineering methodology, was unusually philosophical and unprecedentedly oriented toward symbolism, did not appear until July 1914. The historical question of how much this paper figured in Wittgenstein's work is probably unanswerable. On the very plausible speculation that it did, we can still wonder how differently things might have turned out for Wittgenstein's work had Buckingham's paper not appeared when it did. Would he have had the moment of insight that day in September 1914?

McGuinness writes that "Events like the picture theory's occurring to him were not, I suspect, as crucial as they have been thought. That theory was principally a way of putting more clearly something he had already grasped." I suspect McGuinness is right that the significance of that moment of insight was in being able to put things in terms of pictures and models and hence to put thoughts he had already had more clearly, not in *changing the direction* of his inquiry. But even if Wittgenstein had worked out the same thoughts anyway, perhaps *some* such moment was needed— *some* new way of putting things. A moment of insight that shows you how to put everything together by means of an analogy to a fully worked-out formalism is no small thing.

Again and again, thinkers who have been plagued by a problem that will not let them be report having suddenly lighted onto an analogy that, when worked out, allowed them to proceed from a stalled position. The value to the reader in being able to re-create these unarticulated analogies should probably not be underestimated, but analogies are extremely difficult to convey successfully to someone who has not already at least glimpsed them. Darwin tried to guide his readers along the intellectual path of discovery he had followed from insights about artificial selection (breeding) to his insight about what natural selection had wrought, with questionable success. His statements meant to establish an analogy were often misunderstood, and he was criticized by even his own supporters for statements he made to explain an analogy. They took them out of context and treated them as stand-alone similes; they told him he was causing people to misunderstand his theory. Perhaps, then, Wittgenstein was being wise rather than ungenerous in merely putting the analogies as metaphors, rather than conscientiously explaining the analogy between models and propositions he glimpsed in the fall of 1914.

§§§§

Inasmuch as it has been practical to do so, I have tried to tell the historical story in the preceding chapters independently of presenting my interpretation of portions of the *Tractatus*. My reason for aiming to do so is that the historical story may illuminate different things in the *Tractatus* for others than it does for me, so I wished to tell it in a way that permits others to extract the historical story without too much inconvenience, to make of the elements of the historical story I present in this book what they will.

I do not think of this historical study of the intellectual background as identifying "influences" on Wittgenstein. Rather, I think of the relationship of the *Tractatus* and Wittgenstein's intellectual milieu more along the lines of the relationship between a sculpture made by using "found" objects and the artist's surroundings while working on such a sculpture. Such a sculptor looks at objects in his surroundings—a pair of bicycle handlebars, a radiator—and, in Picasso's case, sees in them the lines of a goat's horns and ribs, something the designers of those objects neither intended nor perceived. Sometimes there is only a superficial similarity of form between the found object and the sculptor's use of it. But sometimes there is more than a mere pun of similar form inspiring the sculptor's use. In Picasso's famous work, the goat horns and the bicycle handlebars are symmetrical along the same axes of motion for the goat as for the bicycle; they are positioned similarly over the goat and bicycle as they move forward, they are swiveled similarly in use, and there are visual and haptic similarities from the standpoint of the person watching or interacting with them. The insight in such metaphors can be helpfully reorientating, just as Wilbur Wright's insight into a bird as an "animated windmill" was.

I think the author of the *Tractatus* made use of things in his intellectual milieu in the way such a sculptor makes use of things in his surroundings. One of these "found" conceptual objects was the very spare and abstract mathematical-logical foundation for the principle of similarity on which the methodology of experimental engineering scale models was based. It was "found" partly because it appeared just when it filled a need. As we have seen, it was presented to the scientific community in 1914, when Wittgenstein was at a resting point in his explorations into logic. He feared he had gone as far as he could go but that it was not far enough. Not long after that he had the crucial insight he claimed to have gained by reflecting on the use of models, which gave him a metaphor for expressing his views about a correct theory of symbolism.

That "found" conceptual object—the very general and philosophically-oriented account of physical similarity by Buckingham heralding the resolution of disagreements between the technical and scientific communities about the validity of experimental engineering models—had been made especially urgent by the increased attention being given to aeronautical research. With encouragement from Russell, Wittgenstein had been decisive about giving up the opportunity for a career in aeronautical research to devote himself to philosophy. But this was not a rejection of the community or values in which he had spent part of his life as an aeronautical researcher; it was a moving on to make the philosophy he had already been working on his main vocation. The *Tractatus* he produced after several more years of work was not the result of winnowing his interests, but of bringing a wide variety of his explorations to bear on the problems he and Russell had run up against in trying to find a correct theory of symbolism: Boltzmann's writings on statistical mechanics and representation in science, Hertz's on the goals of providing a new foundation of physics, Frege's on providing a language for the calculus of thought, for sure, but lots of other things, too. If my argument in this book is correct, Wittgenstein drew on later insights into the methods of flight research to bring all these insights together in an account of how it is possible to speak about things in the world, when we do.

Appendix

On Aeronautics[1]

By Ludwig Boltzmann

Translated by Marco Mertens and Inga Pollmann
(From the German: "Uber Luftschifffarht" in Boltzmann, L. (1905)
Populäre Schriften. Leipzig: J. A. Barth, pp. 81– 91)

On the occasion of the publication of his famous axiom on the division of the circle,[2] Gauß describes—not without pride—how perhaps hundreds of mathematicians had attempted to solve this problem in vain since the time of the ancient Greeks, until, finally, it was considered unsolvable. The same holds true to an even greater degree for the problem of the dirigible.[3] I understand by this every device by means of which one or more persons are able to move a relatively long distance freely through the air in an arbitrary direction.

The number of failed projects in this area is legion. But, starting with the legendary Dedalos and with Leonardo da Vinci, the most outstanding minds have been occupied with this problem throughout the centuries. In fact, there is hardly another problem as alluring for man. Everybody is familiar with the wealth of forms in the world of birds and insects, a wealth that is explained by zoologists as a consequence of their superiority and their ability to spread by virtue of the highly developed flying capacity which these animal classes possess. Now, man, whose railway outstrips the fastest race-horse, whose ships, in spite of their enormous size, scoff at the fish's art of swimming in their maneuverability and mobility on and in water, should he never be able to follow the bird into the sky?

A description of the dirigible's advantages cannot be my task here. I only remark that as the mobility of the means increases, so does the difficulty of using it, yet so does the attainable speed when this difficulty has been overcome. I remember wondering as a child, why, instead of breaking through the isthmus of Suez, one did not rather connect Europe, Asia, and Africa by railroad. I did not yet understand the greater mobility of the ship in water. What advantages, then, shall air—enormously mobile and, additionally, unlimited—have to offer!

It is hardly disputable that the dirigible would bring an upswing to transport, compared to which the one brought about by railroad and steamship would scarcely fall into account. Today's army would not confront the flying-machine—adamantine, sweeping down invulnerably, and throwing dynamite—any differently than a Roman army would have confronted the breech-loaders. Customs would either have to undergo unimagined changes, or cease entirely.

But, in the same way the solution of the problem of the division of the circle (Kreisteilung) failed before Gauß, the construction of the dirigible failed, so that the problem became seriously discredited. Yes, great theoreticians were actually inclined to believe that its solution was impossible. Only recently a turn occurred. The inaccuracy of the old formulas has clearly been shown, and I believe I can provide a proof that the solution to this problem is not only possible but will in all likelihood be found shortly.

From me, as a theoretician, you would most likely expect a long proof founded on sophisticated formulas. But I can do nothing but admit the powerlessness of theoretical mechanics when facing the complex eddies of air. The brevity of time available forbids an exhaustive account of the history of the problem, or an account of the technical details of particular flying devices. Rather, I want to understand the task of theory in a broader sense according to which it has to state the guiding ideas, and to uncover fundamental terms.

The first step towards a solution of the tasks of aeronautics occurred through the invention of the balloon. The major credit for this invention goes to the French, who, at that time, proved themselves a flighty nation in the most favorable sense of the word. The brothers Montgolfier constructed the first balloon filled with hot air; Charles followed soon thereafter with a balloon containing hydrogen. So the first big step had been taken: For the first time, a human being had succeeded in lifting himself freely up into the sky. Only the balloon lacked maneuverability; it was a plaything of the wind.

By now, countless efforts to steer the balloon have followed. One tried to accomplish this by bucket wheels and airscrews, both borrowed from the motion mechanism of the steamship; nor were apparatuses built according to the principle of the motion of skyrockets missing either. In order to push the balloon forward more easily, it was given the shape of a cigar with a pointed tip. Such a balloon, constructed by the French officers Kreß and Renard, moved by means of airscrews and could be steered so well that it truly was a dirigible in absolute calm. Yet the attainable speed fell vastly short of a moderate wind, so that it became a prey to even a light wind.

Indeed, in order to lift a man into the air, a balloon has to have about a thousand times the volume of that same man; in order to carry the specific heavier machine parts, an even greater volume is necessary. But using such gigantic bodies is in direct contradiction to the major property that should characterize the dirigible: light maneuverability. By using a balloon, rapid locomotion is impossible. Even so, the merit of these aeronauts who actually lifted themselves up into the air for the first time cannot be overestimated; their apparatus is still providing excellent services for scientific, military, and other purposes.

But this was only the first step towards the invention of the dirigible. That dirigibles could employ rapid motion—which is indispensable in order to overcome the wind—in order to carry a load, can be seen in the birds of prey, which, after obtaining high speed, keep floating almost without a flap of their wings. Thus we arrive at flying-machines which do not use the aerodynamic lift of a gas that is specifically lighter than air, but rather merely the living power of a mechanism to carry the load in the air. These are called dynamic flying-machines.

They fall into two main categories. In the first category, the motive power is preferably used to lift the device. For this purpose, usually one or more airscrews are used which revolve vertically upwards, just like the screw of a steamship revolves horizontally in water. As in that case, a small part of the whole surface of the screw—two or four uniformly inclined planes that revolve forward in rapid rotation by virtue of their inclination—suffices. The model of this apparatus is a well-known children's toy.

Imagine two or four of such enormous airscrews which are turned very rapidly by a machine: if mounted on a heavy object, it can be lifted up in the air, and you have the helicopter.

In contrast to that, in the second class of the dynamic flying machines, the gliders or aeroplanes, the motive power is mainly used for horizontal motion. Lift-off comes about—according to the most accurate measuring principle followed by Wellner und Lilienthal—due to using a plane with small inclination and curvature which is then . . . drawn through the air at rapid speed. We will call it the principle of the inclined plane. This principle can also be exemplified by a well known toy—the kite. The same constitutes a large, slightly vaulted, and—because of an adherent tail—slightly inclined plane. If pulled quickly through the air on a twine, it ascends to a significant height. The same principle applies to the flight of particularly large birds when they—as already mentioned—float without a beat of their wings in the air after gaining significant speed, which is called gliding. The necessary horizontal movement could be given to the aeroplane either by a kind of wing beat, in which case it would resemble a bird completely, or by the airscrews we already know, which now do not turn vertically, but rather in a horizontal direction.

Mr. Kreß has had the courtesy to provide me with a little model of an apparatus that he had already invented fourteen years ago. He will set the same in motion in front of your eyes, and in doing so illustrate the principle better than would many words. With a problem that difficult, the most thinkable simplification is of paramount importance. Since the horizontal movement for any other flying apparatus has to be produced with similar means, the aeroplane represents the simplest flying-machine imaginable which is able to bear the load capacity without any new mechanism. It essentially imitates the apparatus proven in the flight of birds of prey, and thus has the highest prospect of success from the start.

Numerous flying-machines have been constructed which essentially combine the basic types mentioned here. Numerous airscrews which together constitute an inclined plane; wheels which direct inclined planes in a circle under corresponding steering; combinations of balloons with dynamic flying apparatuses; etc., etc. I am, of course, far from being able to make a disparaging judgment on all these apparatuses, nor do I intend to do so. Still, I am convinced that they have lesser prospects of success because of their greater complexity. Experience also seems to confirm this. At the natural scientists' meeting, which was held in Oxford the previous week, a huge flying-machine constructed by Hiram Maxim—which essentially is a realization in colossal dimensions of the model by Mr. Kreß, just demonstrated—has been the subject of extensive debates. The two airscrews are driven by a most sensibly constructed gas-heated steam engine. The whole flying-machine, which weighs 8,000 English pounds including the two men who operate it, and which moves at a speed of 30 meters per second and therefore faster than the rapidest express train, has indeed been lifted up in the air once. Mr. Maxim has decidedly made the second big step towards the invention of the dirigible. He has shown that one is indeed able to lift huge loads up into the air by means of a dynamic flying apparatus. The greatest English physicists, who are all theoreticians—Lord Kelvin, Lord Rayleigh, Lodge, etc.—spoke of Maxim's machine with excitement, and I already thought that once again, the English would call a new groundbreaking invention their own.

But there is still a catch in this case. Initially, the Maximian machine ran like a locomotive on the rails below it, but when it had reached the necessary speed, it then ran on rails that were mounted above it especially for this purpose. Because of the large buoyant force, one of the upper rails broke too early, and the machine rose into the air; but all of its numerous steering devices could not be launched fast enough; it had to be brought to a halt as quickly as possible, and suffered significant damages. The greatest impediment to all these attempts lies in their dangerous nature. Annoyed, Maxim remarked in his speech that the flight-artist has to be not only a technician, but also an acrobat. Imagine such a big plane moving so fast that its air resistance amounts to about 10,000 pounds, and judge what disturbance each gust, each air swirl will cause the freely floating apparatus which is without a point of support; how colossally each change of inclination, each tilt must affect the motion of the whole. Study the variety and the elegance of the motion of the wings of the bird of prey, consider how a kite turns somersaults in the sky at the slightest carelessness, and imagine the situation of the aeronaut, whose flying-machine loses control in a similar manner.

Surely, now that proof has been provided that the power of the aeroplane suffices to lift huge loads up into the air, it is only a question of skill to steer it properly. Whoever has seen the security with which an enormous ocean liner is steered by a few men; whoever has seen in an ironworks the display of skill, often reproduced, of a steam hammer of 1,000 hundredweights standing still a few millimeters above the crystal of a pocket watch as if acting on order—he will not doubt that one will be able to steer the flying-machine as soon as the necessary experience is gained. But how to gain it without risking human lives?

Would we dare, after merely theoretical explanation of the machinery, to allow even the most intelligent men to steer an ocean liner through dangerous cliffs if they have never seen a ship? And others, in that case, had already tried out the machines earlier. Thus, we would almost be tempted, despite the ingenious achievements of Maxim, to apply a trivial Berlin proverb[4] to his apparatus.

Every invention has its preworkers and its subsequent improvers. But still, one man almost always has to be designated as the actual inventor. Who, now, will be the actual inventor of the dirigible? Today, Maxim is not yet that one. Only he who is actually able to fly in any direction at will, with and against the wind, so long as a relatively large fuel quantity lasts (about one hour) will be the actual inventor of the dirigible.

This invention has not been made yet, nor has the time yet arrived to outrun the English. Admittedly, we cannot achieve it through generosity of capital. Maxim's machine is reported to have cost 300,000 florins. But how much the Germans already have accomplished by means of splendid ideas, though with minimal capital! Who would dispute this here in Vienna, where *Die Zauberflöte*, the *9th Symphony*, and the *Missa Solemnis* have been written? They should try to imitate this throughout the entire rest of the world if they can.

Of course, I do not mean by that that all great Germans of the future should be no more praised than Mozart. Not every man is as eternally sanguine as he was; not every field of activity is as independent as music. For lack of assistance, Ressel had to leave the entire benefit and half of the glory of his invention to the English. By contrast, then, I should like to propose at the business meeting of our young society of natural scientists—which is likewise still balancing somewhat in the air—to make their first grant benefit aeronautics, or, if their capital does not suffice, to prompt governments for such support.

Otto Lilienthal, engineer in Berlin, has succeeded in an experiment, which I would like to designate as the third step towards the invention of the dirigible. Shipping on water did not begin with the ocean liner, but with a tub built from a hollowed trunk. Likewise, Lilienthal began with a flying apparatus that was as small as possible. He equipped his arms with two wings—initially tightly connected—of 15 square meters surface area, which in their essentials imitated the wings of a bird. These wings represent an aeroplane that is capable of carrying a man at sufficient speed. To obtain this speed, Lilienthal dispensed with any motor; he simply ran a distance against the wind, and then jumped—resting upon his wings—into the air. Of course, because he had no power source, he could not fly arbitrarily far, and he could fly upwards only to an extremely limited measure. But by means of initially fairly short, and later longer, jumps, he eventually succeeded in floating a distance of 250 meters over a slightly inclined slope on the Rhinow Mountain, keeping himself rather close to the earth the whole time. There he was assured of the danger of being overturned or being tilted sharply by a gust of wind, but also the possibility of acquiring absolute mastery at steering through yearlong practice. Lilienthal accomplishes this steering by shifting the body and by foot movements using a steering wheel that imitates the tail of a bird, but which is nevertheless fixed.

Lilienthal now intends to carry a very small motor. As he increases the power of that motor he hopes to be able to adjust the size of the wings and the achieved skillfulness in steering to the new circumstances, until the horizontal locomotion obtained by the motor suffices to keep the flying person constantly above ground. Admittedly, this flying apparatus can have very little practical significance at present. Vast improvements and a much larger scale of construction would be necessary before the economic and social consequences discussed above would result. However, the problem would be solved theoretically, a way leading to the goal found, and the actual invention of the dirigible performed. The theoretical discovery of the proper way often precedes its perfection in practical usage. Did the first telegraphs, the first photographs already have a practical use, or would the discovery of America have had such economic consequences if the journey was still as trying for us as it had been for Columbus?

I have yet to mention that Kreß has contrived a very promising steering apparatus based on different principles, although not yet tested at bigger loads.

With respect to the apparatus used for the generation of horizontal motion, there are also divergent views. All of the technological mechanisms create a so-called cyclical motion; i.e., a motion in which all constituent parts will again arrive at their initial position after a shorter or longer amount of time. There are two systems of cyclical motion, the rotating and the reciprocating. The different wheels and inductors of dynamos are examples of the first; the pistons of steam engines and pumps of the latter. In the case of locomotion in water by means of bucket wheels, the first system is applied; in the case of oars and the fins of fish, the second. In the case of flying, Lilienthal prefers the second system, which is also employed in nature in the flight of birds, whereas the first system, the application of airscrews for the generation of horizontal motion, has no counterpart in nature. It must be acknowledged that in the construction of acoustical and optical apparatuses, and of pumps and locomotion mechanisms, animal organs always serve as models only up to a certain limit, because nature works with differing means and pursues differing ends. Even though rotating apparatuses are almost completely alien to nature, our bucket wheels and water-screws work successfully in place of the reciprocating fins, and velocipedes work in place of literally reciprocating feet.

According to Lilienthal, the aeroplane has to be divided into two halves, both of which move like the wings of birds in flight. As a result, both the sliding (the so-called slip[5]) of the screws and the loss of power that results from the production of eddies will be avoided, and Lilienthal believes that as a result, one loses less power to the air. Yet I actually doubt this, because much of the work achieved by bringing down the wings gets lost in lifting them again, whereas the very profitable principle of the inclined plane can be optimally applied to the airscrew. Maxim's airscrews, in fact, operate with very little slip. By contrast, the division of the aeroplane into two wings affects its firmness and simplicity considerably. One cannot attain flapping of the wings without considerable complication and significant friction of the mechanism, and it does not operate as continuously as the airscrew, nor is it as sharply adjustable. In addition, to calculate the effect of the flapping wing beforehand is far more difficult.

As a result, an aeroplane driven by airscrews appears to be the theoretically most promising mechanism, and the only one which actually has already been lifted into the air in the case of both small models and larger exemplars.

It is unbelievable how simple and natural any result appears once it has been found, but how challenging, as long as the way that leads us there is unknown. In the same way, steering an aeroplane will at some point be performed by craftsmen with ease, but it can be discovered only by a genius of first grade. And this inventor has to be not only a genius, but also a hero. One cannot wring its secrets with little effort from the element that has to be newly conquered. Only he who possesses the courage to trust his life to the new element and the cunning to overcome gradually all its treacheries has a chance to kill the dragon which, until this very day, deprives mankind of the treasure of this invention. The inventor of the dirigible has to resemble in this way the archetype of all grand discoverers, Christopher Columbus, by personal courage as well as by ingenuity, has set an example to all discoverers of the future. "Who great things would win, he must dare to die."[6] As a result, should someone—not taught by our century's countless wonders of technology—scoff at the attempts to fly, we will take to heart the words with which the most idealist poet addressed the greatest discoverer:

> Steer on, bold sailor—Wit may mock thy soul that sees the land,
> And hopeless at the helm may droop the weak and weary hand,
> Yet ever—ever to the West, for there the coast must lie,
> And dim it dawns, and glimmering dawns before thy reason's eye;
> Yet trust thy guiding god—and go along the floating grave,
> Though hid till now—Yet now behold the New World o'er the wave!
> With genius Nature ever stands in solemn union still,
> And ever what the one foretells the other shall fulfill.[7]

Besides deliberation and ardor, only one thing is necessary, which even Columbus attained only with the utmost difficulty: money.

---

1 Talk held at the Gesellschaft Deutscher Naturforscher und Ärzte in Vienna, 1894.

2 Translator's note: also known as Gauß' 5th theorem. *See Karl Friedrich Gauß, Disquistiones Arithmeticae, XX. Edition, Yale University Press, New Haven, 1966, first published in 1801.*

3 „lenkbares Luftschiff" in the original version.

4 Perhaps "Jedes Wort ist ein verlorener Kuss" ("Every word is a missed kiss," the implication being that some things can be learned only through experience).

5 English in the original version.

6 Translator's note: quote from Friedrich Schiller, *Reiterlied* (The song of the Horsemen), 1797, as well as *Wallenstein. Ein dramatisches Gedicht*, completed 1797. *1. Teil: Wallensteins Lager* (The camp of Wallenstein), *11. Aufzug*: „Und setzet ihr nicht das Leben ein, /Nie wird euch das Leben gewonnen sein" (Who life would win, he must dare to die). Boltzmann changed the meaning by substituting „Leben" with „Großes."

7 Translator's note: Friedrich Schiller, *Columbus*, 1795.

Notes

These notes are not referenced to superscript numbers in the main text. Rather, the initial number in each note refers to a page number. The phrase in quotes following the page number appears on that page. The ensuing note refers to that phrase on that page.

Preface

vii **"glimpsed the 'fundamental thought' (or *Grundgedanke*, in his native German)":** Wittgenstein, Ludwig. G. H. von Wright and G. E. M. Anscombe (eds.) (1984). *Notebooks: 1914–1916*, 2nd ed. Chicago: University of Chicago Press. On p. 37e (the entry for 25 December 1914) he says, "My fundamental thought is that logical constants are not proxies." On p. 7e (the entry for 29 September 1914), upon thinking about scale models, he glimpses that "The general concept of the proposition carries with it a quite general concept of co-ordination of proposition and situation," which is the idea he uses to eliminate the need for logical constants to be proxies. My attention was brought to this and other texts about the fundamental thought of the *Tractatus* being that logical constants are not proxies by Brian McGuinness's essay "The *Grundgedanke* of the *Tractatus*" in McGuinness, Brian, *Approaches to Wittgenstein: Collected Papers*. New York and London, Routledge. 103–115.

vii **"he was confident it contained the solution to all the problems of philosophy":** Wittgenstein, Ludwig. D. F. Pears and B. F. McGuinness (eds.) (1972). *Tractatus Logico-Philosophicus*, reprinted with corrections. New York: Humanities Press. 5.

viii **"Edgar Buckingham's 'On Physically Similar Systems'":** Buckingham, Edgar (1914). "On Physically Similar Systems: Illustrations of the use of dimensional equations." *Physical Review*, 4, 345–376.

viii   **"he left Manchester for Cambridge in 1911"**: Stated in many sources, including McGuinness, Brian F. (1988), *Wittgenstein: A Life: Young Ludwig 1889–1921*, Berkeley: University of California Press, 73. (Oxford University Press issued a reprint of this Wittgenstein biography with a new preface in 2005, entitled *Young Ludwig: Wittgenstein's Life 1889–1921.*)

ix   **"a possible state of affairs"**: Von Wright, G. H. (n.d.). *Wittgenstein.* Minneapolis: University of Minnesota Press. 21.

ix   **"Wittgenstein's own patented propeller design"**: I surveyed a variety of biographical sources around 1993 and then performed an extensive literature survey in 1998–9. The six suggestions I have listed can be found in, collectively: Janik, Allan and Stephen Toulmin (1973), *Wittgenstein's Vienna*, New York: Simon and Schuster; McGuinness (1988), cited above; Monk, Ray (1990), *Wittgenstein: The Duty of Genius*, New York: Penguin Books; Brockhaus, Richard R. (1991), *Pulling Up the Ladder*, La Salle: Open Court; and Nedo, Michael (1993), *Wittgenstein: Introduction, Weiner Ausgabe/Vienna Edition*, New York: Springer Verlag; among others.

The most well-known of the six suggested influences is the first, Hertz's *Principles of Mechanics*, which is developed most fully in Griffin, James (1964), *Wittgenstein's Logical Atomism*, Oxford: Clarendon Press, and is discussed at length in Chapters 5 and 6 of Janik and Toulmin (1973) and Chapter 7 of Brockhaus (1991). The explicit suggestion that the idea of a proposition as a picture owes much to the study of descriptive geometry and graphical statics studied at the Technische Hochschule is made in McGuinness (1988) on page 61.

As part of my survey of the relevant literature, I also attended a talk in 1998 later published as Hamilton, Kelly (2001). "Wittgenstein and the Mind's Eye." In Klagge, James C. (ed.) (2001). *Wittgenstein: Biography and Philosophy*. Cambridge, UK: Cambridge University Press. In her talk, Hamilton discussed the relevance of points in Eugene S. Ferguson's *Engineering and the Mind's Eye* (MIT Press, 1992) vis-à-vis quotes from the *Tractatus*. Hamilton presented some of what was already previously known about Wittgenstein's education and added new details about Wittgenstein's course work at the Technische Hochschule, where he studied engineering. She emphasized visual thinking as the way to understand the notions of model and picturing in the *Tractatus*, in addition to emphasizing geometric projection (which was suggested previously by McGuinness). Hamilton's view presented there (which was prior to 2000) is in *the completely opposite*

direction of my view that the notion of a model in the *Tractatus* is an experimental engineering scale model. As Boltzmann emphasized, such visual
thinking is insensitive to scale, and experimental engineering models are
not. Boltzmann's article on models distinguished the two kinds of models in
terms of precisely this difference. Hamilton's discussion did not mention the
methodology of experimental engineering scale models, nor did it mention
dimensionless parameters or the principle of dynamic similarity.

In my survey in the late 1990s, I did learn of one young philosopher working on the notion of scale models as cognitive representations: Jonathan
Waskan, who shared some parts of his (then-unpublished) dissertation with
me; he has since published on the topic in Waskan, J. A. (2003), "Intrinsic
Cognitive Models," *Cognitive Science* 27/1. 259–283. Though interesting in
its own right, and a promising contribution to the field of cognitive science
and philosophy of psychology, his work did not discuss dimensionless
parameters, which was the facet of the methodology I planned to empha-
size. Nor was his work particularly related to Wittgenstein's *Tractatus*.

Thus my statement that, prior to presenting my work, no one had ever suggested that the methodology of experimental engineering scale models was
reflected in the *Tractatus*. It was only after the publication of the abstract of
my "Physical Pictures" talk in early 2000, presentation of my "Physical
Pictures" talks in 2000, and distribution of the texts and handouts associated with those talks that the notions of dimensionless parameters, the
methodology of experimental engineering scale models, and Buckingham's
writings on physically similar systems entered the Wittgenstein literature.

x   **"when I constructed a timeline"**: A timeline was provided on handouts distributed at the talk "Physical Pictures: Engineering models circa 1914 and in
Wittgenstein's *Tractatus*," presented in July 2000 in Vienna, Austria at the
History of Philosophy of Science meeting. A revised version of the timeline
was also distributed at a longer talk of the same name, given at the
University of North Carolina at Chapel Hill in November 2000. It may be
found on page 22 of the free pdf file at http://philsci-archive.pitt.edu/
archive/00000661.

x   **"Buckingham had studied for his doctorate under Ostwald"**: Stock, John
T. (2003). *Ostwald's American Students: Apparatus, Techniques, and Careers.*
Concord, NH: Plaidswede Publishing. 135.

xi   **"it would make practical manned heavier-than-air powered flight possible"**: There are many excellent historical accounts of the Wright Brothers'

invention of the airplane. Here and in the following paragraphs on the Wright Brothers, I consulted: Crouch, Tom (1989), *The Bishop's Boys: A Life of Wilbur and Orville Wright*, New York: WW Norton; Tobin, James (2003), *To Conquer the Air: The Wright Brothers and the Great Race for Flight*, New York: Free Press; Heppenheimer, T. A. (2003), *First Flight: The Wright Brothers and the Invention of the Airplane*, New York: John Wiley & Sons, Inc.; Hallion, Richard P. (2003), *Taking Flight: Inventing the Aerial Age from Antiquity Through the First World War*, New York: Oxford University Press.

xiv   **"to develop similar research facilities in the U.S.":** These accounts come from books by NASA historians (National Aeronautics and Space Administration, formerly NACA, National Advisory Committee on Aeronautics). Successive accounts are given in: Anderson, Frank W. Jr. (1976), *Orders of Magnitude: A History of NACA and NASA, 1915–1976* (NASA SP-4403), Washington, D.C.: Scientific and Technical Information Office; Roland, Alex (1985), *Model Research: The National Advisory Committee for Aeronautics 1915–1958*, Volume 1, Washington, D.C.: Scientific and Technical Information Office; Gorn, Michael H. (2001), *Expanding the Envelope: Flight Research at NACA and NASA*, Lexington, KY: University Press of Kentucky. See also Baals, Donald D. and William R. Corliss (1981), *Wind Tunnels of NASA* (NASA SP-440), Washington, D.C.: Scientific and Technical Information Office.

xiv   **"had written a book on the foundations of thermodynamics":** Buckingham, Edgar (1900). *An Outline of the Theory of Thermodynamics.* New York: The Macmillan Co.; London: Macmillan and Co., Ltd.

xiv   **"arose sometime after 1911":** In Buckingham, Edgar (1921). "Notes on the Theory of Dimensions." *Philosophical Magazine* v. 42, No. 251 (November 1921). 696n.

xiv   **"which appeared first in July 1914":** Buckingham, Edgar (1914a). "On Physically Similar Systems." *Journal of the Washington Academy of the Sciences*, July 1914.

xvi   **"his thinking had been influenced by both of these physicists' works":** McGuinness (1988). Ibid. 37.

xvi   **"encapsulated in what he called a 'theory of types'":** Russell, Bertrand (1908). "Mathematical Logic as Based on the Theory of Types." *American Journal of Mathematics* 30. 222–262.

xvi **"It took place in September 1914"**: Wittgenstein, Ludwig. G. H. von Wright and G. E. M. Anscombe (eds.) (1984). *Notebooks: 1914–1916*, 2nd ed. Chicago: University of Chicago Press. 7e.

xvii **"a motor-car accident is represented by means of dolls, etc."**: Wittgenstein, Ludwig. G. H. von Wright and G. E. M. Anscombe (eds.) (1984). *Notebooks: 1914–1916*, 2nd ed. Chicago: University of Chicago Press. 7e.

xviii **"They rejected it"**: Recounted in von Wright (n.d.). *Wittgenstein.* Minneapolis: University of Minnesota Press. 95.

xviii **"he wrote back to Wrinch"**: Recounted in von Wright (n.d.). Ibid. 95.

xx **"he was displeased with the editorial changes made"**: Recounted in von Wright (n.d.). Ibid. 100.

Chapter 1: Toys to Overcome Time, Distance, and Gravity

1 **"was born near"**: McGuinness, Brian F. (2005). *Young Ludwig: Wittgenstein's Life 1889–1921*. Oxford University Press. viii. (In the new preface to the 2005 edition, McGuinness corrects the statement in the original edition with new information that indicates that the house in which Wittgenstein was born was not at that time within the city of Vienna.)

1 **"an engineer designing new steel mills"**: McGuinness, Brian F. (1988). *Wittgenstein: A Life: Young Ludwig 1889–1921*. Berkeley: University of California Press. 12–14. Also in Nedo, Michael (1993). *Wittgenstein: Introduction. Weiner Ausgabe/Vienna Edition*. New York: Springer Verlag. 12.

1 **"Gustav Mahler, and many other composers"**: Nedo, Michael (1993). *Wittgenstein: Introduction. Weiner Ausgabe/Vienna Edition*. New York: Springer Verlag. 13.

2 **"is now available as an MP3 file on the Internet"**: As of the time of this writing (March 11, 2005), the web site http://www.measure.demon.co.uk/sounds/Brahms.html provides a link from which to download an MP3 version of the recording made by Brahms.

2 **"a musical composition other than by hearing a live performance"**: There was, in addition, the representation of a piano performance on a punched paper roll. But this was not a general means of recording music, since it was

a method specifically for compositions performed on a piano. A player piano roll from a live piano performance was produced that could then be played back mechanically on a player piano, reproducing the original performance to some extent.

2   **"lying around open as one wandered about":** Flindell, E. (1971). *Music Review* 32. Quoted in Michael Nedo, op. cit., 13.

3   **"re-heard in every well-furnished parlor?":** Berliner, Emile (1888). "The Gramophone: Etching the Human Voice." *Journal of the Franklin Institute of the State of Pennsylvania*, Vol. CXXV., No. 6. 446.

4   **"the long Allegasse analyses that followed each Vienna Philharmonic Concert":** McGuinness (1988), op. cit. 21.

4   **"the relationship between 'language and the world'":** Wittgenstein, Ludwig. D. F. Pears and B. F. McGuinness (eds.) (1972). *Tractatus Logico-Philosophicus*, reprinted with corrections. New York: Humanities Press. 4.014.

4   **"They are all constructed according to a common logical pattern":** Ibid.

5   **"projects the symphony into the language of musical notation":** Ibid.

6   **"into the language of gramophone records":** Ibid. 4.0141.

6   **"the visual record of sound as wavy lines":** My reflections on the processes involved in sound recording were stimulated by reading a version of Biggs, Michael (2004), "Visualisation and Wittgenstein's *Tractatus* in Malcolm, M. (ed.), *Multidisciplinary Approaches to Visual Representations and Interpretations*, Amsterdam: Elsevier B.V., 2004, 293–303, which was available online in 2002. In particular, Biggs considers the information that is contained in a DVD recording of a symphony performance versus other kinds of recordings and representations of it.

7   **"the well-known phonautograph of Leon Scott":** Houston, Edwin J. (1888). "The Gramophone." *Journal of the Franklin Institute*, January 1888.

7   **"imploring drag of the organ-grinder's tuneful melody":** Berliner, Emile (1890). "The Improved Gramophone." A paper read at the 52nd meeting of the American Institute of Electrical Engineers, New York, December 16. 21.

8   **"graphically with a view to industrial applications"**: Marty, Daniel (2004). "Leon Scott and His Phonautograph." *The History of the Phonautograph.* http://www.phonautograph.com. Version updated January 15, 2004. This site contains many articles about, and sketches and photographs of, the machine, and explains its operation.

8   **"as a means of conveying the gramophone sound recording"**: Berliner, Emile (1890). "The Improved Gramophone." A paper read at the 52nd meeting of the American Institute of Electrical Engineers, New York, December 16. 23.

8   **"thousands of miles away"**: Ibid. 28.

9   **"from the line on a gramophone record"**: Wittgenstein, Ludwig. D. F. Pears and B. F. McGuinness (eds.) (1972). *Tractatus Logico-Philosophicus,* reprinted with corrections. New York: Humanities Press. 4.0141.

9   **"which makes it into a picture"**: Ibid. 2.1514.

10   **"the remarkable volume of the Count's voice"**: Morton, Frederic (1979). *A Nervous Splendor: Vienna 1888/1889.* New York: Penguin Books. 49–50.

10   **"was never heard of again"**: Ibid. 50.

11   **"from which he wanted to extract everything"**: McGuinness (1988). op. cit. 34.

11   **"referred to as 'the father of aeronautics'"** Hallion, Richard P. (2003). *Taking Flight: Inventing the Aerial Age from Antiquity Through the First World War.* New York: Oxford University Press. 105.

11   **"to become interested in flight"**: Gibbs-Smith, Charles H. (1966). *The Invention of the Aeroplane, 1799–1909.* London: Faber & Faber. 19.

12   **"obtained with screws"**: Chanute, Octave (1998). *Progress in Flying Machines* (originally published in 1894). New York: Dover Publications. 55–56.

12   **"a perfect lightweight powerplant"**: Hallion, Richard P. (2003). *Taking Flight: Inventing the Aerial Age from Antiquity Through the First World War.* New York: Oxford University Press. 121.

12   **"before he had reached 30 years of age"**: Chanute (1998). Ibid. 122.

12   **"both disabling and painful"**: Gibbs-Smith, Charles H. (1966). *The Invention of the Aeroplane, 1799–1909*. London: Faber & Faber. 21. Hallion, Richard P. (2003). *Taking Flight: Inventing the Aerial Age from Antiquity Through the First World War*. New York: Oxford University Press. 120–121.

13   **"in 11 seconds"**: Penaud, Alphonse "Aéroplane Automoteur: Équilibre Automatique," L'aéronaute, v. 5, n. 1 (January 1872): 5–6; see also 2–9 and figures 3 and 4, quoted in Hallion (2003). Ibid. 122.

13   **"was the talk of the aeronautical world"**: Hallion (2003). Ibid. 121–122.

14   **"Can the operator control it?"**: Hallion (2003). Ibid. 121–122.

16   **"compendious expositions of a new logic"**: Janik, Allan and Stephen Toulimin (1973). *Wittgenstein's Vienna*. New York: Simon and Schuster. 107.

17   **"Wittgenstein wanted to design, build, and fly his own airplane"**: McGuinness (1988). Ibid. 69.

Chapter 2: To Fly Like a Bird, Not Float Like a Cloud

19   **"committed suicide years before"**: Hallion (2003). Ibid. 125.

19   **"was first published"**: Lilienthal, Otto (1889). *Der Vogelflug als Grundlage der Fliegekunst (Bird Flight as a Basis of Aviation)*. Berlin: (unknown). The second edition was published in 1910 with an additional chapter by Gustav Lilienthal.

20   **"flew it spectacularly well"**: Hallion (2003). Ibid. 230–233.

20   **"flying kites that summer"**: McGuinness (1988). Ibid. 64.

21   **"successfully flew it for a quarter-mile"**: Jenkins, Garry (2001). *Colonel Cody and the Flying Cathedral: The Adventures of the Cowboy Who Conquered the Sky*. New York: Picador. 165.

21   **"swung him 'like a clock pendulum'"**: Jenkins (2001). Ibid. 44.

22   **"words like 'tarnation' and 'geossifax' abounded"**: Jenkins (2001). Ibid. 76.

22   **"without stirring a wing"**: Jenkins (2001). Ibid. 82–83.

22   **"each problem as it arose"**: Jenkins (2001). Ibid. 85.

22   **"trial and error, test, fail, try again"**: Jenkins (2001). Ibid. 86.

23 **"to outdo them all":** Jenkins (2001). Ibid. 82.

23 **"an instrument of war":** Jenkins (2001). Ibid. 87.

23 **"she went aloft in the Cody family's invention herself":** Jenkins (2001). Ibid. 181.

24 **"relieved of his position":** Jenkins (2001). Ibid.

24 **"over 50,000 people attended his funeral":** The Drachen Foundation (2004). "The Cody Collection." http://www.drachen.org/about_archive _cody.html. Accessed August 4, 2004. There is much contradictory material on Cody, the confusion caused largely by Cody's own intentional deceptions to portray himself as from the "Wild West" and to cause people to confuse him with the original "Buffalo Bill," a name he used in his earlier endeavors on the stage. The Drachen Foundation has recently acquired many original archival materials on Samuel F. Cody and promises to be a good resource, as the foundation is making them available for free public use. Garry Jenkins' biography of Cody, cited above, is a good resource on Cody as well.

25 **"research and development":** Jenkins (2001). Ibid. 238.

25 **"before Lilienthal's book appeared":** Martgot, Jean-Michel and Zvi Har'El (2004). "The Complete Jules Verne Bibliography." http://jv.gilead.org.il/ biblio/. Accessed July 10, 2004. Section I.

25 **"a proposed powered balloon for air travel":** Robert Wohl's 1994 book, *A Passion for Wings: Aviation and the Western Imagination, 1908 to 1918*, New Haven: Yale University Press, convinced me of the relevance of Verne's novel predicting the superiority of heavier-than-air flight to understanding the cultural and intellectual atmosphere existing prior to the invention of the airplane.

26 **"is a practical matter":** Verne, Jules (2004). *Robur the Conqueror*. (English translation of *Robur le Conquerant*.) Free Public Domain E-Books from the Classic Library. http://www.jules-verne.co.uk/robur-the-conqueror/. Accessed February 3, 2004. 9.

26 **"has never constructed anything flying":** Verne, Jules (2004). *Robur the Conqueror*. (English translation of *Robur le Conquerant*.) Free Public Domain E-Books from the Classic Library. http://www.jules-verne.co.uk/robur-the-conqueror/. Accessed March 10, 2005. 10.

27   **"according to Dr. Marcy, of the Institute of France":** Verne, Jules (2004). Ibid. 19-20.

27   **"Lilienthal's first known lecture on aviation":** "Otto Lilienthal's Aeronautical Bibliography: An Annotated Overview." Otto Lilienthal Museum web site. http://www.lilienthal-museum.de/olma/e4.htm. Accessed March 10, 2005.

27   **"the shape of linens hanging in a breeze":** Lilienthal, Otto (2001). *Birdflight as the Basis of Aviation: A Contribution Towards a System of Aviation.* Translated from the second edition by A. W. Isenthal in 1911. Hummelstown, PA: Markowski International Publishers. 88.

27   **"if the restraining crossbar is removed":** Lilienthal, Otto (2001). Ibid. 92.

28   **"large enough to carry a human":** Harper, Harry (1942). *Man's Conquest of the Air.* London: Scientific Book Club. 21.

28   **"Gustav recounted in an introduction":** Lilienthal, Otto (2001). Ibid. xii.

28   **"Gustav and Otto Lilienthal joined the British one":** "Otto Lilienthal's Aeronautical Bibliography: An Annotated Overview." Ibid.

29   **"moved forward at the same level":** Lilienthal, Otto (2001). Ibid. ix–xx.

29   **"with only negative results":** Lilienthal, Otto (2001). Ibid. xxi.

29   **"could not accompany Otto to the sandhill":** Lilienthal, Otto (2001). Ibid. xxiii.

30   **"normally used as a safety measure":** Lilienthal, Otto (2001). Ibid. xxiii–xiv.

31   **"to transfer our highway to the air":** Lilienthal, Otto (2001). Ibid. 1–2.

32   **"what nature demonstrates to us daily in birdflight":** Lilienthal, Otto (2001). Ibid. 2.

32   **"through it in a horizontal direction":** Lilienthal, Otto (2001). Ibid. 5.

32   **"would completely solve the mystery of flight":** Lilienthal, Otto (2001). Ibid. 50.

33   **"the development of mechanical flight":** Lilienthal, Otto (2001). Ibid. 106.

33   **"to solve the great problem of aerial navigation":** Lilienthal, Otto (2001). Ibid. 106.

33 **"practical and free navigation of the air"**: Lilienthal, Otto (2001). Ibid. 107.

33 **"the direction in which the wind blows"**: Lilienthal, Otto (2001). Ibid. 108.

34 **"nothing came of it at first"**: Hallion, Richard P. (2003). *Taking Flight: Inventing the Aerial Age from Antiquity Through the First World War.* New York: Oxford University Press. 48.

34 **"astonishing the onlookers"**: Hallion, Richard P. (2003). Ibid. 49.

34 **"French and British scientific establishment"**: Hallion, Richard P. (2003). Ibid. 49.

34 **"to keep the balloon aloft"**: Hallion, Richard P. (2003). Ibid. 49.

35 **"balloon over Paris"**: Hallion, Richard P. (2003). Ibid. 49.

35 **"a repetitive pattern of fleur-de-lis"**: Hallion, Richard P. (2003). Ibid. 53.

35 **"It had happened so fast"**: Hallion, Richard P. (2003). Ibid. 57.

36 **"depicted on dinnerware"**: Hallion, Richard P. (2003). Ibid. 58.

36 **"you may think proper to say"**: Hallion, Richard P. (2003). Ibid. 59.

36 **"balloons were developed for military use"**: Hallion, Richard P. (2003). Ibid. U.S. Centennial of Flight Commission (2003). "Military Use of Balloons in the Mid- and Late Nineteenth Century." Chapter 4. http://www.centennialofflight.gov/essay/Lighter_than_air/military_balloons_in_Europe/LTA4.htm. Accessed March 10, 2005. Among many others.

36 **"back over the Austrian troops"**: U.S. Centennial of Flight Commission (2003). Ibid.

37 **"forbidden indulgence of time and attention"**: Tobin, James (2003). Ibid. 65.

37 **"he described his preliminary model experiments"**: Maxim, Hiram (1891). "The Maxim Flying Machine." *Manufacturer and Builder.*

38 **"will be successful where I failed"**: Maxim, Hiram (1891). "The Maxim Flying Machine." *Manufacturer and Builder.* 127.

38 **"he was working on going fast"**: Maxim, Hiram (189?). "A New Flying Machine." 444. Maxim, Hiram (1895?). "Aerial Navigation."

39 **"to show others how not to fly"**: McCallum, Iain (1999). *Blood Brothers.* London: Chatham Publishing. 111.

40   **"we should have a veritable flying machine"**: Maxim, Hiram S. "The Development of Aerial Navigation." First page.

40   **"now we have to keep it chained"**: Maxim, Hiram S. 444.

41   **"the subject of an entire book"**: McCallum, Iain (1999). *Blood Brothers*. London: Chatham Publishing.

41   **"on the topic"**: Boltzmann, Ludwig (1894, 1905). *Uber Luftschifffarht*. In Boltzmann, L. (1905). *Populare Schriften*. Leipzig: J. A. Barth. 81–91. Other Wittgenstein biographers (such as Nedo and McGuinness) have noted this lecture of Boltzmann's in connection with Wittgenstein's aeronautical career.

43   **"similar in their smallest parts"**: O'Conner, J. J. and E. F. Robertson (1996). "Johann Carl Friedrich Gauss." http://www-history.mcs.st-andrews.ac.uk/Mathematicians/Gauss.html. Accessed August 6, 2005.

44   **"the only known Fermat primes"**: Weisstein, Eric W. "Fermat Prime." From *MathWorld*—A Wolfram Web Resource. http://mathworld.wolfram.com/FermatPrime.html. Accessed March 11, 2005.

44   **"be found shortly"**: Boltzmann, L. "On Aeronautics." Translated by Marco Mertens and Inga Pollmann (from the German: "Uber Luftschifffarht" in Boltzmann, L. (1905), *Populare Schriften*.) Leipzig: J. A. Barth. 81–91. Appendix to this book.

46   **"well-known children's toy"**: Ibid.

46   **"you have the helicopter"**: Ibid.

46   **"a well-known toy—the kite"**: Ibid.

46   **"referred to Maxim's machine with ardor"**: Ibid.

47   **"of this invention"**: Ibid.

Chapter 3: Finding a Place in the World

50   **"he should continue in philosophy"**: McGuinness, Brian F. (1988). *Wittgenstein: A Life: Young Ludwig 1889–1921*. Berkeley, University of California Press. 92.

51   **"was a compendium of information"**: Chanute, Octave (1998). *Progress in Flying Machines*. (Originally published in 1894.) Mineola, New York: Dover Publications.

51   **"the predictions they expected"**: Crouch, Tom (1989). *The Bishop's Boys: A Life of Wilbur and Orville Wright*. New York: WW Norton. 212.

51   **"had a plan for a flying machine"**: Crouch, Tom (1989). Ibid. 253.

51   **"in 1903 in France"**: Gibbs-Smith, Charles H. (1974). *The Rebirth of European Aviation*. 56, 71–81.

52   **"the rest of the world"**: Bancroft, Hubert Howe (1893). *World's Columbian Exposition of 1893: Book of the Fair*. Chicago: The Bancroft Company. http://columbus.gl.iit.edu. Accessed on March 12, 2005.

52   **"'Only in Paris … !'"**: Bancroft, Hubert Howe (1893). Ibid. 17. ("Said American visitors to the Paris Fair: Only in Paris can such marvels be accomplished!")

52   **"magnificent grouping of buildings"**: Bancroft, Hubert Howe (1893). Ibid.

53   **"or chemical apparatus"**: Bancroft, Hubert Howe (1893). Ibid. 308.

53   **"the bicycle exhibits"**: Crouch, Tom (1989). Ibid. 151.

54   **"only in 1887"**: Crouch, Tom (1989). Ibid. 106.

54   **"a peak of 1.2 million by 1895"**: Crouch, Tom (1989). Ibid. 106.

54   **"was irresistible"**: Crouch, Tom (1989). Ibid. 107.

54   **"in need of bicycle repairs"**: Crouch, Tom (1989). Ibid. 107.

55   **"entrance into Yale College"**: Crouch, Tom (1989). Ibid. 74.

55   **"a game on skates"**: Crouch, Tom (1989). Ibid. 74. (Quoted from Milton Wright's diary.)

55   **"to store for future use"**: Crouch, Tom (1989). Ibid. 7. (Quoted from Milton Wright's diary.)

55   **"and self doubt"**: Crouch, Tom (1989). Ibid. 76.

56   **"Times were hard"**: Crouch, Tom (1989). Ibid. 90.

56   **"to larger jobs"**: Crouch, Tom (1989). Ibid. 95.

56   **"in March 1889"**: Crouch, Tom (1989). Ibid. 96.

57 **"and Machinery Buildings":** "World's Columbian Exhibition—Reactions." http://xroads.virginia.edu/~MA96/WCE/reactions.html. Accessed February 23, 2005.

57 **"they had repaired and sold":** Crouch, Tom (1989). Ibid. 111.

57 **"made to order":** Crouch, Tom (1989). Ibid. 113.

57 **"had begun to wander":** Crouch, Tom (1989). Ibid. 114.

58 **"to fly with his brother":** Crouch, Tom (1989). Ibid. 57.

58 **"serious attention to the bicycle":** Crouch, Tom (1989). Ibid. 115.

58 **"citing three events":** Crouch, Tom (1989). Ibid. Chapter 10.

59 **"and therefore safer, aircraft":** Crouch, Tom (1989). Ibid. 169.

59 **"an animated windmill":** Crouch, Tom (1989). Ibid. 172.

60 **"it was possible to fly":** Based on Fred C. Kelly reminiscence: "I recall one other question, about when he got the biggest 'kick' out of the invention. Was it when the machine took off in the first flight ever made? 'No,' Orville said, I got more thrill out of flying before I had ever been in the air at all— while lying in bed thinking how exciting it would be to fly," in Kelly, Fred C., Editor, *How We Invented the Airplane: An Illustrated History by Orville Wright.* New York: Dover Publications, Inc. 55.

61 **"they were not the first":** Gibbs-Smith, Charles H. *The Aeroplane: An Historical Survey of its Origins and Development.* London: Her Majesty's Stationery Office. 226.

62 **"data had not been wrong":** Heppenheimer, T. A. (2003). *First Flight: The Wright Brothers and the Invention of the Airplane.* New York: John Wiley & Sons, Inc. 149–151. Also explained in Tobin, James (2003). *To Conquer the Air: The Wright Brothers and the Great Race for Flight.* New York: Free Press. 129.

64 **"had done what they said":** Tobin, James (2003). Ibid. 224.

64 **"to gain a contract":** Jenkins, Garry (2001). *Colonel Cody and the Flying Cathedral: The Adventures of the Cowboy Who Conquered the Sky.* New York: Picador. 151–152.

64 **"Cody had derided":** Jenkins, Garry (2001). Ibid. 135–136.

65   **"in making short turns"**: Orville Wright (1929). "Wilbur Wright by Orville Wright." *Encyclopedia Brittanica Online*, 14th edition.

65   **"a market for the invention"**: Orville Wright (1929). Ibid.

65   **"a much more unified affair"**: Sopka, Katherine P. (ed.). *Physics for a New Century: Papers Presented at the 1904 St. Louis Congress*. Introduction by Albert E. Moyer. Tomash Publishers/American Institute of Physics. ix.

66   **"words, equations, or symbols"**: Sopka, Katherine P. (ed.). Ibid. xvi.

66   **"of the American delegation"**: Sopka, Katherine P. (ed.). Ibid. xvii.

66   **"use of the 'dimensional formula'"**: Sopka, Katherine P. (ed.). Ibid. xix.

66   **"to that of principles"**: Langevin, Paul (1904). "The Relations of Physics of Electrons to Other Branches of Science." In Sopka, Katherine P. (ed.). Ibid. 195.

66   **"variety of initial conditions"**: Boltzmann, Ludwig (1904). "The Relations of Applied Mathematics." In Sopka, Katherine P. (ed.). Ibid. 277.

66   **"a senseless and hopeless manner"**: Boltzmann, Ludwig (1904). Ibid. 276.

67   **"fond of quoting"**: McGuinness, Brian (1988). Ibid. 39.

67   **"endowed with certain properties"**: Boltzmann, Ludwig (1904). "The Relations of Applied Mathematics." In Sopka, Katherine P. (ed.). Ibid. 276.

67   **"a mathematical-physical theory"**: Boltzmann, Ludwig (1904). Ibid. 276.

69   **"account of the Congress"**: Davis, William H. (1904). "International Congress of Arts and Sciences." *Popular Science Monthly*. 66.

69   **"as 'a formal rebuttal'"**: Sopka, Katherine P. (ed.). Ibid. xvi.

70   **"multiplied by a distance"**: Nichols, Edward Leamington (1904). "The Fundamental Concepts of Physical Science." In Sopka, Katherine P. (ed.). Ibid. 95.

70   **"definite knowledge of physics"**: Nichols, Edward Leamington (1904). Ibid. 96.

71   **"from its axioms and definitions"**: Nichols, Edward Leamington (1904). Ibid. 96.

71   **"of the mechanics involved"**: Nichols, Edward Leamington (1904). Ibid. 96.

71   **"by its very nature is arbitrary":** Fourier (1888). "Analytical Theory of Heat." Quoted in Palacios, J. *Dimensional Analysis*. London: Macmillan & Co., Ltd. ix.

72   **"when Wittgenstein was five years old":** McGuinness, Brian F. (1988). *Wittgenstein: A Life: Young Ludwig 1889–1921*. Berkeley: University of California Press. 46.

72   **"as a natural environment":** McGuinness, Brian F. (1988). Ibid. 25.

72   **"in which they lived":** McGuinness, Brian F. (1988). Ibid. 25.

73   **"his harshness towards Hans":** McGuinness, Brian F. (1988). Ibid. 27.

73   **"regarded in the family":** McGuinness, Brian F. (1988). Ibid. 26.

73   **"criticism and envy":** McGuinness, Brian F. (1988). Ibid. 17.

73   **"I could see nothing against it":** McGuinness, Brian F. (1988). Ibid. 48.

74   **"to study science":** McGuinness, Brian F. (1988). Ibid. 27, 38.

74   **"as a great joke":** McGuinness, Brian F. (1988). Ibid. 28.

74   **"impending (because wished for) death":** McGuinness, Brian F. (1988). Ibid. 50.

74   **"would prepare him for":** McGuinness, Brian F. (1988). Ibid. 50.

74   **"constructing a model sewing machine":** McGuinness, Brian F. (1988). Ibid. 45.

74   **"or much happiness":** McGuinness, Brian F. (1988). Ibid. 50.

74   **"of Ludwig's childhood":** McGuinness, Brian F. (1988). Ibid. 47.

74   **"go to the theatre":** Nedo, Michael (1993). *Wittgenstein: Introduction. Weiner Ausgabe/Vienna Edition*. New York: Springer Verlag. 14.

75   **"was in physics":** McGuinness, Brian F. (1988). Ibid. 51. Also, that Wittgenstein listed physics as his chosen career is reported in Spelt, P. D. M. and B. F. McGuinness (2000). "Marginalia in Wittgenstein's copy of Lamb's *Hydrodynamics*." *Wittgenstein Studies 2*. 131–148. (Published as *From the Tractatus to the Tractatus and Other Essays*. G. Oliveri (ed.). Peter Lang. Frankfurt. 2001.)

75 **"solved the problem of flight":** Gibbs-Smith, Charles H. (1974). The Rise of European Aviation: 1902–1908. London: Her Majesty's Stationery Office. 189.

75 **"projects connected with it":** McGuinness, Brian F. (1988). Ibid. 55.

75 **"of his divided vocation":** McGuinness, Brian F. (1988). Ibid. 54.

76 **"winter semester of 1906–1907":** Eftekhari, Ali (2003). *Ludwig Boltzmann (1844–1906).* http://philsci-archive.pitt.edu/archive/00001717/. 6.

76 **"the best of the German engineering schools":** McGuinness, Brian F. (1988). Ibid. 54.

77 **"McGuinness's suggestion":** McGuinness, Brian F. (1988). Ibid. 61.

78 **"have coauthored an article":** Spelt, P. D. M. and B. F. McGuinness (2000). "Marginalia in Wittgenstein's copy of Lamb's *Hydrodynamics.*" *Wittgenstein Studies 2.* 131–148. (Published as *From the Tractatus to the Tractatus and Other Essays.* G. Oliveri (ed.). Peter Lang. Frankfurt. 2001.)

78 **"his invention":** There are many excellent photographs of kites designed and built by Cody. My favorite is the centerfold of Garry Jenkins' book on Cody, cited above, showing the normally overbearing Cody looking diminutive, contemplative, even childlike, in a huge room surrounded by man-carrying kites of his own invention. There are many online sources as well. As of this writing, the web site http://www.hsa.lr.tudelft.nl/~frits/cody.html contains many beautiful photographs of kites carrying humans, a kite fitted with a basket for meteorological instruments, and the man-lifting system.

78 **"on its maiden flight":** Jenkins, Garry (2001). Ibid. 140.

79 **"the historical development of aeronautics":** Hide, Oystein (2004). "Wittgenstein's Books at the Bertrand Russell Archives and the Influence of Scientific Literature on Wittgenstein's Early Philosophy," *Philosophical Investigations* 27:1 68–91 lists the following related to ballooning and early aeronautics: Bourgeois, M. David, Rechereches sur L'art de voler, Cuchet, Paris 1784; Turbini, D. Gasparo, La nuova Scoperta del Globo Aerostatico di Montgolfier, Brescia 1784; other 1784 works in which the author's name is unknown are described on pp. 79 and 85 of Hide (2004); a 1789 work in Italian on aeronautics, especially ballooning; an 1891 edition of a six-volume work containing the manuscripts of Leonardo da Vinci; a 1908 French work about "activities in the air" (Berget, A. Ballons, dirigeables et ae'ro-planes, Libraire Universalle, Paris 1908); and a three-volume work in

German, published in Leipzig in 1821, with accounts of various balloon flights. Hide mentions that the book collection also contains two major works by Frege—*Begriffsschrift: eine der Arithmeticschen nachgebildete FORMELSPRACHE des reinen Denkens*, published in 1879, and his *Grundlagen der Arithmetik, Eine logisch mathematische Untersuchung uber den Begriff der Zahl*, published in 1884—that are thought to have belonged to Wittgenstein; these might well have been purchased later. Thus, I have not included them among the books he is thought to have acquired while at the THS.

79   **"published in London in 1730":** Hide, Oystein (2004). Ibid. 84.

79   **"fond of quoting from Hertz":** McGuinness, Brian (1988). Ibid. 39.

80   **"authors read in the Allegasse":** McGuinness, Brian (1988). Ibid. 39.

80   **"with the family with whom he stayed":** McGuinness, Brian (1988). Ibid. 57–60.

Chapter 4: A New Continent

81   **"his own kite research":** McGuinness, Brian (1988). Ibid. 64.

81   **"how to make a kite":** McGuinness, Brian (1988). Ibid. 64.

81   **"the whole project":** McGuinness, Brian (1988). Ibid. 64.

82   **"all aspects of the subject":** McGuinness, Brian (1988). Ibid. 65.

82   **"became Professor of Engineering":** McGuinness, Brian (1988). Ibid. 64.

82   **"had expressed interest in studying":** McGuinness, Brian (1988). Ibid. 62.

82   **"with the kites":** McGuinness, Brian (1988). Ibid. 64.

82   **"out of Wittgenstein's work":** McGuinness, Brian (1988). Ibid. 66.

83   **"of storms":** McGuinness, Brian (1988). Ibid. 66.

83   **"concerts in Manchester":** McGuinness, Brian (1988). Ibid. 66.

83   **"for engine design":** McGuinness, Brian (1988). Ibid. 70.

83   **"commanded respect and high salaries":** Harper, Harry (1929*). Twenty-five Years of Flying.* London: Hutchinson & Co. 137.

83   **"with Bertrand Russell":** McGuinness, Brian (1988). Ibid. 73.

84   **"'enormous' personal income":** McGuinness, Brian (1988). Ibid. 71.

84   **"and apparently expensively":** McGuinness, Brian (1988). Ibid. 71.

84   **"talk about music":** McGuinness, Brian (1988). Ibid. 71.

85   **"to discontinue experiments":** Letter from Wilbur and Orville Wright on Wright Cycle Company letterhead to Mr. Georges Besancon, dated November 17, 1905. Quoted in Gibbs-Smith, Charles H. (1974). Ibid. 177–179.

85   **"the veracity of the Wright Brothers":** Gibbs-Smith, Charles H. (1974). Ibid. 179.

85   **"the success of their experiments":** Gibbs-Smith, Charles H. (1974). Ibid. 179.

85   **"the British *Automotor Journal*":** Gibbs-Smith, Charles H. (1974). Ibid. 195.

86   **"'But we must hurry up'":** "The Great French Manifesto: August 1906." Quoted in Gibbs-Smith, Charles H. (1974). Ibid. 213–215.

87   **"doubted his achievements":** Gibbs-Smith, Charles H. (1974). Ibid. 280.

87   **"would be childish":** Gibbs-Smith, Charles H. (1974). Ibid. 280.

89   **"for propeller designs for airships":** National Archives [of the UK], Department of Scientific and Industrial Research (DSIR). DSIR 23/7.

90   **"to prepare for the tripos":** R Glazebrook, Horace Lamb, *Obituary Notices of Fellows of the Royal Society of London 1* (1935), 375-392.

92   **"had discussed similar questions":** McGuinness, Brian (1988). Ibid. 74.

92   **"(who had 'solved' Russell's contradiction)":** McGuinness, Brian (1988). Ibid. 74.

92   **"Russell's *Principles of Mathematics*":** McGuinness, Brian (1988). Ibid. 75.

92   **"their definition of number":** McGuinness, Brian (1988). Ibid. 75.

93   **"of an experiment":** McGuinness, Brian (1988). Ibid. 73.

93   **"some 'objections to his theories'":** Entry for 1911, date unknown, in Wittgenstein Gesamtbriefwechsel INTELEX. Accessed August 7, 2005.

Chapter 5: A New Age-Old Problem to Solve

95   **"become a philosopher"**: McGuinness, Brian F. (1988), *Wittgenstein: A Life: Young Ludwig 1889–1921*. Berkeley: University of California Press. 92n.

95   **"unless he is some good"**: McGuinness, Brian (1988). Ibid. 92.

95   **"to help me judge"**: McGuinness, Brian (1988). Ibid. 92.

95   **"given him encouragement"**: McGuinness, Brian (1988). Ibid. 93.

95   **"the great moments of his life"**: McGuinness, Brian (1988). Ibid. 115.

96   **"they learnt 40 or 50 songs"**: McGuinness, Brian (1988). Ibid. 123.

96   **"at the time"**: McGuinness, Brian (1988). Ibid. 125.

96   **"in 'cottage rooms'"**: McGuinness, Brian (1988). Ibid. 125.

96   **"musical perceptions in 1905"**: Faculty of Music, University of Cambridge online catalog entry. http://www.mus.cam.ac.uk/external/research/science-andmusic.html. Accessed March 8, 2005.

96   **"thriving today"**: Science & Music Group, Faculty of Music, University of Cambridge. http://www.mus.cam.ac.uk/~ic108/SM/SM.html. Accessed March 8, 2005.

97   **"about the work there"**: Hindshaw, W. (1938). Reminiscences. Samuel Alexander Papers, John Rylands Library, University of Manchester. Quoted in Costall, Alan. "Pear and his peers: the beginnings of psychology at Manchester." http://www.psy.man.ac.uk/history/history.pdf. Accessed March 9, 2005.

97   **"the psychology of music"**: Costall, Alan. Ibid. 9.

97   **"to be a subject"**: McGuinness, Brian (1988). Ibid. 125.

98   **"and probably attended"**: McGuinness, Brian (1988). Ibid. 127.

98   **"as we know it"**: McGuinness, Brian (1988). Ibid. 126.

98   **"were unable to hear it as music"**: McGuinness, Brian (1988). Ibid. 126.

98   **"of reality as we imagine it"**: Wittgenstein, Ludwig (1974). *Tractatus Logico-Philosophicus*. Translated by D. F. Pears and B. F. McGuinness. 4.01.

98   **"of what they represent"**: Wittgenstein, Ludwig (1974). Ibid. 4.011.

99   **"to Russell as 'absurd'":** McGuinness, Brian (1988). Ibid. 128.

99   **"the philosophy of psychology":** Wittgenstein, Ludwig (1974). Ibid. 4.1121.

99   **"there is an analogous risk":** Wittgenstein, Ludwig (1974). Ibid. 4.1121.

99   **"worked with the mathematician Philip Jourdain":** Nedo, Michael (1993). *Wittgenstein: Introduction. Weiner Ausgabe/Vienna Edition*. New York: Springer Verlag. 18.

100   **"there are no logical constants":** Letter to Bertrand Russell, June 22, 1912. Wittgenstein Gesamtbriefwechsel, INTELEX. Accessed August 7, 2005.

100   **"Frege called these 'pseudo-propositions'":** In Hilbert-Frege correspondence, cited and discussed in Sterrett (1994), "Frege and Hilbert on the Foundations of Geometry." Available online at http://philsci-archive.pitt.edu/archive/00000723/.

101   **"there were no logical constants":** It is striking that he mentions this in the same letter as Myers and James, for William James' *Principles of Psychology* discusses the role of connectives in language, and Wittgenstein later specifically addressed James's remarks about "a feeling of and, a feeling of if, a feeling of but." Henry Jackman's discussion in "Wittgenstein's and James's 'Stream of Thought'" is one of the fairest treatments of it I have read. (Jackman, Henry. "Wittgenstein's and James's 'Stream of Thought.'" Paper presented at the Society for the Advancement of American Philosophy, 31st Annual Conference, Radisson Hotel Birmingham, March 4–6, 2004.)

101   **"I now write R (a,b)":** Wittgenstein (1974). Letters to Russell, Keynes, and Moore. Oxford: Basil Blackwell. Letter to Russell #9 dated January 1913.

101   **"but it is true!":** Wittgenstein Gesamtbriefwechsel. Letter to Bertrand Russell, August 19, 1919. Electronic INTELEX Past Masters Database. Accessed August 7, 2005.

102   **"aims at effecting":** "Theory of Logical Types" in Russell, B. *Essays in Analysis*. New York: Braziller 1973. 216.

103   **"the hierarchy of types":** Ibid. 216–217.

104   **"who looked to Wittgenstein's judgement":** As evident by the progression of their relationship chronicled in Chapter 5 of McGuinness (1988). Ibid.; see especially page 176.

Chapter 6: The Physics of Miniature Worlds

105   **"these properties which it must have":** Wittgenstein, Ludwig. "Notes Dictated to G. E. Moore April 1914." Appendix II to Wittgenstein, Ludwig (1979). *Notebooks 1914–1916*. Edited by G. H. von Wright and G. E. M. Anscombe. With an English translation by G. E. M. Anscombe. Chicago: University of Chicago Press. 107.

105   **"the way in which they correspond":** Wittgenstein, Ludwig. "Notes Dictated to G. E. Moore April 1914." Appendix II to Wittgenstein, Ludwig (1979). *Notebooks 1914–1916*. Edited by G. H. von Wright and G. E. M. Anscombe. With an English translation by G. E. M. Anscombe. Chicago: University of Chicago Press. 113.

105   **"the form of a proposition":** Wittgenstein, Ludwig. "Notes on Logic." Appendix I to Wittgenstein, Ludwig (1979). *Notebooks 1914–1916*. Edited by G. H. von Wright and G. E. M. Anscombe. With an English translation by G. E. M. Anscombe. Chicago: University of Chicago Press. 105.

105   **"having those relations in reality":** Wittgenstein, Ludwig. "Notes Dictated to G. E. Moore April 1914." Appendix II to Wittgenstein, Ludwig (1979). *Notebooks 1914–1916*. Edited by G. H. von Wright and G. E. M. Anscombe. With an English translation by G. E. M. Anscombe. Chicago: University of Chicago Press. 113.

106   **"what each word signifies":** Wittgenstein, Ludwig. "Notes on Logic." Appendix I to Wittgenstein, Ludwig (1979). *Notebooks 1914–1916*. Edited by G. H. von Wright and G. E. M. Anscombe. With an English translation by G. E. M. Anscombe. Chicago: University of Chicago Press. 100.

106   **"a manifold of dimensions":** Boltzmann, Ludwig, B. F. McGuinness (ed.) (1974). *Theoretical Physics and Philosophical Problems: Selected Writings (Vienna Circle Collection)*. Boston, MA: D. Reidel. 226.

106   **"definite areas of fact":** Boltzmann, Ludwig, B. F. McGuinness (ed.) (1974). *Theoretical Physics and Philosophical Problems: Selected Writings (Vienna Circle Collection)*. Boston, MA: D. Reidel. 226.

107   **"to study physics with him":** McGuinness, Brian (1988). Ibid. 54.

107   **"can be predicted":** Boltzmann, Ludwig, B. F. McGuinness (ed.) (1974). *Theoretical Physics and Philosophical Problems: Selected Writings (Vienna Circle Collection)*. Boston, MA: D. Reidel. 42.

107   **"only a mental picture"**: Boltzmann, Ludwig, B. F. McGuinness (ed.)
(1974). *Theoretical Physics and Philosophical Problems: Selected Writings
(Vienna Circle Collection)*. Boston, MA: D. Reidel. 42.

107   **"the wood for the trees"**: Boltzmann, Ludwig, B. F. McGuinness (ed.)
(1974). *Theoretical Physics and Philosophical Problems: Selected Writings
(Vienna Circle Collection)*. Boston, MA: D. Reidel. 43. A number of commentators have pointed out the relevance of these and similar passages to
Wittgenstein's early thoughts: McGuinness, cited above; Janik and Toulmin,
cited above; and Peter Barker (in Barker, Peter [1980], "Hertz and
Wittgenstein," *Studies in the History and Philosophy of Science*, 11: 243–256).

107   **"those which appear in nature"**: Boltzmann, Ludwig, B. F. McGuinness
(ed.) (1974). *Theoretical Physics and Philosophical Problems: Selected
Writings (Vienna Circle Collection)*. Boston, MA: D. Reidel. 214.

108   **"is treated in the article UNITS, DIMENSIONS OF"**: Boltzmann, Ludwig,
B. F. McGuinness (ed.) (1974). *Theoretical Physics and Philosophical
Problems: Selected Writings (Vienna Circle Collection)*. Boston, MA: D.
Reidel. 220.

108   **"with a toy airplane"**: Spelt, P. D. M. and B. F. McGuinness (2000).
"Marginalia in Wittgenstein's copy of Lamb's *Hydrodynamics*." *Wittgenstein
Studies 2*, 131–148. (Published as *From the Tractatus to the Tractatus and
Other Essays*. G. Oliveri (ed.). Peter Lang. Frankfurt. 2001.)

110   **"Wenham had convinced"**: Baals, Donald D. and William R. Corliss (1981).
Wind Tunnels of NASA (NASA SP-440). Washington, D.C.: Scientific and
Technical Information Office. 3.

112   **"that I sit in this chair"**: Wittgenstein, Ludwig. "Notes on Logic." Appendix
I to Wittgenstein, Ludwig (1979). *Notebooks 1914–1916*. Edited by G. H. von
Wright and G. E. M. Anscombe. With an English translation by G. E. M.
Anscombe. Chicago: University of Chicago Press. 97.

112   **"associated with a spoken sentence"**: Russell, Bertrand and Alfred North
Whitehead (1997). *Principia Mathematica*, 2nd ed. Cambridge, England:
Cambridge University Press. 402.

113   **"to this new method"**: Boltzmann, Ludwig, B. F. McGuinness (ed.) (1974).
*Theoretical Physics and Philosophical Problems: Selected Writings (Vienna
Circle Collection)*. Boston, MA: D. Reidel. 11.

113   **"of real numbers in 1872":** O'Connor, J. J. and E. F. Robertson (1998). Georg Ferdinand Ludwig Philipp Cantor. http://www-history.mcs. st-andrews.ac.uk/Mathematicians/Cantor.html.

114   **"points on the line":** Dedekind, Richard (1963). "Continuity and Irrational Numbers." In *Essays on the Theory of Numbers.* New York: Dover Publications. 1–24.

116   **"a paper about similar structures":** Thompson, James (1875). "Comparison of similar structures as to elasticity, strength, and stability." In Thompson, James (1912). *Collected papers in physics and engineering,* 1st ed. 1912. 361–372.

117   **"homologous linear dimensions":** Thompson, James (1875). Ibid. 362.

117   **"lecture he gave in 1911":** Thompson, D'Arcy Wentworth (1911). "Magnalia Naturae: of the Greater Problems of Biology." In *Science*, October 6, 1911. New Series, Vol. 34, No. 875, 417–428.

118   **"alongside our own":** Thompson, D'Arcy Wentworth (1911). Ibid. 419.

118   **"by its relation to mathematics":** Thompson, D'Arcy Wentworth (1911). Ibid. 426.

118   **"in its relation to mathematics":** Thompson, D'Arcy Wentworth (1992). *On Growth and Form,* complete revised edition. New York: Dover Publications. 1.

118   **"visions of Plato and Pythagoras":** Thompson, D'Arcy Wentworth (1992). Ibid. 2.

118   **"between the apple and the stars":** Thompson, D'Arcy Wentworth (1992). Ibid. 9.

119   **"been pursued by few":** Thompson, D'Arcy Wentworth (1992). Ibid. 9.

119   **"the selfsame laws":** Thompson, D'Arcy Wentworth (1992). Ibid. 11.

119   **"evil betrays itself in another":** Thompson, D'Arcy Wentworth (1992). Ibid. 13.

119   **"all its work is done":** Thompson, D'Arcy Wentworth (1992). Ibid. 42.

119   **"depend on gliding more and more":** Thompson, D'Arcy Wentworth (1992). Ibid. 42.

119   **"to compare one bird with another"**: Thompson, D'Arcy Wentworth (1992). Ibid. 44.

119   **"is little worth the making"**: Thompson, D'Arcy Wentworth (1992). Ibid. 44.

120   **"fast enough through the air"**: Thompson, D'Arcy Wentworth (1992). Ibid. 43.

120   **"at which its flight is stable"**: Thompson, D'Arcy Wentworth (1992). Ibid. 45.

120   **"as its size increases"**: Thompson, D'Arcy Wentworth (1992). Ibid. 41.

120   **"their support and stability"**: Thompson, D'Arcy Wentworth (1992). Ibid. 46.

120   **"the case of man"**: Thompson, D'Arcy Wentworth (1992). Ibid. 47–8.

120   **"essential to the large"**: Thompson, D'Arcy Wentworth (1992). Ibid. 48.

120   **"add power to their glider"**: Thompson, D'Arcy Wentworth (1992). Ibid. 48–9.

121   **"of cement and steel"**: Thompson, D'Arcy Wentworth (1992). Ibid. 28.

121   **"its whole environment or milieu"**: Thompson, D'Arcy Wentworth (1992). Ibid. 24.

122   **"save the least of organisms"**: Thompson, D'Arcy Wentworth (1992). Ibid. 51.

122   **"which they may attain"**: Thompson, D'Arcy Wentworth (1992). Ibid. 51.

122   **"an inefficient and inappropriate mechanism"**: Thompson, D'Arcy Wentworth (1992). Ibid. 52.

122   **"the force of surface tension"**: Thompson, D'Arcy Wentworth (1992). Ibid. 57.

122   **"have acted upon it"**: Thompson, D'Arcy Wentworth (1992). Ibid. 16.

123   **"that acts upon matter"**: Thompson, D'Arcy Wentworth (1992). Ibid. 15.

123   **"in connection therewith"**: Thompson, D'Arcy Wentworth (1992). Ibid. 16.

123   **"or against breadth"**: Thompson, D'Arcy Wentworth (1992). Ibid. 78.

124   **"rate of growth"**: Thompson, D'Arcy Wentworth (1992). Ibid. 78.

124 **"in various directions":** Thompson, D'Arcy Wentworth (1992). Ibid. 79.

124 **"pure mathematical reasoning":** Thompson, D'Arcy Wentworth (1992). Ibid. 17n.

124 **"to apply":** Thompson, D'Arcy Wentworth (1992). Ibid. 15.

125 **"to be considered":** Thompson, D'Arcy Wentworth (1992). Ibid. 81.

126 **"play their several roles":** Thompson, D'Arcy Wentworth (1992). Ibid. 77.

126 **"must be recast":** Thompson, D'Arcy Wentworth (1992). Ibid. 77.

127 **"the principle of similitude":** Thompson, D'Arcy Wentworth (1992). Ibid. 79.

128 **"instance of 'dimensional theory'":** Thompson, D'Arcy Wentworth (1992). Ibid. 31.

128 **"cites with approval":** Thompson, D'Arcy Wentworth (1992). Ibid. 31n.

129 **"are of great importance":** Thompson, D'Arcy Wentworth (1992). Ibid. 25.

129 **"the volume of the body":** Thompson, D'Arcy Wentworth (1992). Ibid. 25.

130 **"an English translation":** Galileo, G. (1914). *The Two New Sciences by Galileo Galilei.* (Tr. by Henry Crew and Alfonso de Salvio.) Evanston, IL: Northwestern University Press.

130 **"by 1914":** Galileo, G. (1914). *The Two New Sciences by Galileo Galilei.* (Tr. by Henry Crew and Alfonso de Salvio.) Evanston, IL: Northwestern University Press. vi.

130 **"in 1665":** Galileo, G. (1914). *The Two New Sciences by Galileo Galilei.* (Tr. by Henry Crew and Alfonso de Salvio.) Evanston, IL: Northwestern University Press. v.

130 **"an imperfect one":** Galileo, G. (1914). *The Two New Sciences by Galileo Galilei.* (Tr. by Henry Crew and Alfonso de Salvio.) Evanston, IL: Northwestern University Press. vi.

130 **"the following translation":** Galileo, G. (1914). Ibid. v.

130 **"Crew lectured":** Later published in *Science*, New Series, Vol. 37, No. 952 (March 28, 1913), 463–470.

131 **"within his scientific field"**: Hide, Oystein (2004). "Wittgenstein's Books at the Bertrand Russell Archives and the Influence of Scientific Literature on Wittgenstein's Early Philosophy." *Philosophical Investigations.* 27:1 74.

132 **"was a physicist"**: American Institute of Physics. "Biography of Henry Crew." In "Finding Aid to the Papers of Henry Crew." http://www.aip.org/history/ead/northwestern_crew/19990056_content.html. Accessed December 3, 2004.

133 **"a manifest error"**: Galileo, G. (1914). *The Two New Sciences by Galileo Galilei.* (Tr. by Henry Crew and Alfonso de Salvio.) Evanston, IL: Northwestern University Press. 52–53.

133 **"the squares of the times"**: Galileo, G. (1914). *The Two New Sciences by Galileo Galilei.* (Tr. by Henry Crew and Alfonso de Salvio.) Evanston, IL: Northwestern University Press. 139.

134 **"length of the string 144 cubits"**: Galileo, G. (1914). *The Two New Sciences by Galileo Galilei.* (Tr. by Henry Crew and Alfonso de Salvio.) Evanston, IL: Northwestern University Press. 140.

134 **"a large number of vibrations"**: Galileo, G. (1914). Ibid. 140.

134 **"about similarity"**: Stanton, T. E. and J. R. Pannell (1914). "Similarity of Motion in Relation to the Surface Friction of Fluids." *Proceedings of the Royal Society of London, Series A, Containing Papers of a Mathematical and Physical Character.* Vol. 90, No. 619 (July 1, 1914). 394–395.

135 **"that of Osborne Reynolds"**: Stanton and Pannell (1914). Ibid. 200.

135 **"Reynolds' major discoveries"**: Stanton and Pannell (1914). Ibid. 200.

136 **"similarity in the motions"**: Stanton and Pannell (1914). Ibid. 201.

136 **"depends on"**: Stanton and Pannell (1914). Ibid. 201.

137 **"an equation of form (A)"**: Rayleigh. *Scientific Papers.* No. 340, "Note as to the Application of the Principle of Dynamical Similarity." 532–533.

137 **"in early 1914"**: Rayleigh (1914). "Fluid Motions." *Scientific Papers.* No. 384, Vol. VI, p. 237. Also in *Nature*, Vol. XCIII, 364. Also *Proceedings of the Royal Institute.* March 1914.

138 **"observed at another"**: Rayleigh (1914). "Fluid Motions." Ibid. 246.

138 **"at the *corresponding* places"**: Rayleigh (1914). "Fluid Motions." Ibid. 246.

138   **"with that of sound"**: Rayleigh (1914). "Fluid Motions." Ibid. 246.

139   **"Rayleigh also pointed out"**: Rayleigh (1914). "Fluid Motions." Ibid. 237.

139   **"the difficulty is greatest"**: Rayleigh (1914). "Fluid Motions." Ibid. 245–246.

140   **"for model-prototype similarity"**: Rouse, Hunter and Simon Ince (1957). *History of Hydraulics*, 1st ed. Iowa City: Institute of Hydraulic Research. 200.

140   **"what Rayleigh was referring to"**: Rayleigh (1914). "Fluid Motions." 246.

140   **"much hesitation in applying it"**: Rayleigh (1914). "Fluid Motions." 246.

141   **"calculations of surface friction"**: Stanton and Pannell (1914). Ibid. 199–200.

141   **"from experiments on models"**: Stanton and Pannell (1914). Ibid. 201.

142   **"Onnes had published a paper"**: Onnes, H. Kamerlingh (1881). Algemene theorie der vloeistoffen, *Verhandelingen Koninklijke Akademie van Wetenschappen*. 21.

142   **"about how molecules behaved"**: "Scientists of the Dutch School" from the "Van der Waals" web site at the Royal Netherlands Academy of Arts and Sciences. http://www.knaw.nl/waals/kamerlingh.html. Accessed in November 2004.

142   **"nothing more about it there"**: van der Waals (1910). "The equation of state for gases and liquids." Nobel lecture, December 12, 1910. 264.

142   **"seemed particularly important"**: Onnes, Kamerlingh (1913). Nobel lecture. 306.

144   **"the idea of continuity occurred to me"**: Onnes, Kamerlingh (1913). Nobel lecture. Ibid. 255.

144   **"introduced by van der Waals"**: Levelt Sengers, Johanna (2002). *How Fluids Unmix: Discoveries by the School of Van der Waals and Kamerlingh Onnes*. Amsterdam: Edita, KNAW.

145   **"for all fluids"**: Levelt Sengers, Johanna (2002). Ibid. 25.

145   **"of the reference fluid"**: Levelt Sengers, Johanna (2002). Ibid. 26.

146   **"and time are changed"**: Levelt Sengers, Johanna (2002). Ibid. 30.

147   **"are in corresponding states"**: Levelt Sengers, Johanna (2002). Ibid. 30.

147   **"an exact replica of that in the first fluid"**: Levelt Sengers, Johanna (2002). Ibid. 30.

147   **"The 'Principle of Similitude'"**: Tolman, Richard C. (1914). "The Principle of Similitude." *Physical Review*. 3 244–255.

147   **"in every respect to the present universe"**: Tolman, Richard C. (1914). "The Principle of Similitude." *Physical Review*. 3 244.

148   **"in the real universe"**: Tolman, Richard C. (1914). "The Principle of Similitude." *Physical Review*. 3 244.

148   **"must measure the same for O and O'"**: Tolman, Richard C. (1914). "The Principle of Similitude." *Physical Review*. 3 245.

149   **"of the quantity in question"**: Tolman, Richard C. (1914). "The Principle of Similitude." *Physical Review*. 3 247.

149   **"principle of the relativity of size"**: Tolman, Richard C. (1914). "The Principle of Similitude." *Physical Review*. 3 255.

150   **"Reynolds' most famous work"**: Reynolds, Osborne (1883). "An Experimental Investigation of the Circumstances Which Determine Whether the Motion of Water Shall be Direct or Sinuous, and of the Law of Resistance in Parallel Channels."

151   **"the answers were not obvious"**: Reynolds, Osborne (1883). "An Experimental Investigation of the Circumstances Which Determine Whether the Motion of Water Shall be Direct or Sinuous, and of the Law of Resistance in Parallel Channels." 937.

151   **"the primary result of it"**: Reynolds, Osborne (1883). "An Experimental Investigation of the Circumstances Which Determine Whether the Motion of Water Shall be Direct or Sinuous, and of the Law of Resistance in Parallel Channels." 935.

151   **"the fluid and the velocity"**: Reynolds, Osborne (1883). "An Experimental Investigation of the Circumstances Which Determine Whether the Motion of Water Shall be Direct or Sinuous, and of the Law of Resistance in Parallel Channels." 935.

152   **"he later authored a textbook on it"**: Tolman, Richard Chace (1938/1980). *The Principles of Statistical Mechanics*. New York: Dover Publications.

152   **"directly determined experimentally":** American Institute of Physics. *Interview with Linus Pauling, Ph.D.* November 11, 1990. Big Sur, California. http://www.achievement.org/autodoc/page/pau0int-1. Accessed November 2004.

Chapter 7: Models of Wings and Models of the World

155   **"from experiments on models":** Stanton, T. E. and J. R. Pannell (1914). "Similarity of Motion in Relation to the Surface Friction of Fluids." *Proceedings of the Royal Society of London, Series A, Containing Papers of a Mathematical and Physical Character.* Vol. 90, No. 619 (July 1, 1914).

155   **"Horace Lamb's *Hydrodynamics*":** Lamb, Horace (1994/1932). *Hydrodynamics,* 6th ed. Originally published in 1892. Cambridge, UK: Cambridge University Press.

155   **"A. H. Gibson's *Hydraulics and Its Applications*":** Gibson, A. H. (1912). *Hydraulics and Its Applications.* London: Constable & Co., Ltd.

156   **"the two different groups":** Jackson, J. D. (n.d.). "Osborne Reynolds: Scientist, Engineer and Pioneer." Manchester School of Engineering web site. http://www.eng.man.ac.uk/historic.reynolds/oreyna.htm. Accessed July 6, 2003.

157   **"all the phenomena investigated":** Reynolds, Osborne (1879). "On Certain Dimensional Properties of Matter in the Gaseous State. Part I. Experimental Researches on Thermal Transpiration of Gases through Porous Plates and on the Laws of Transpiration and Impulsion, Including an Experimental Proof that Gas is Not a Continuous Plenum. Part II. On an Extension of the Dynamical Theory of Gas, Which Includes the Stresses, Tangential and Normal, Caused by a Varying Condition of Gas, and Affords an Explanation of the Phenomena of Transpiration and Impulsion." *Philosophical Transactions of the Royal Society of London,* Vol. 170, 842.

158   **"of every disregarded factor":** Gibson, A. H. (1912). Ibid. From Preface to 1908 edition. (no pg. no.)

159   **"structure of hydraulics":** Gibson, A. H. (1912). Ibid. From Preface to 1908 edition. (no pg. no.)

159 **"present the experimental data"**: Stanton, T. E. and J. R. Pannell (1914). "Similarity of Motion in Relation to the Surface Friction of Fluids." *Proceedings of the Royal Society of London, Series A, Containing Papers of a Mathematical and Physical Character*. Vol. 90, No. 619 (July 1, 1914). 201.

160 **"the presentation of his data"**: Anderson, J. D., Jr. (1997). *A History of Aerodynamics*. Cambridge, UK and New York: Cambridge University Press. 224.

160 **"in affiliation with the National Physical Laboratory"**: Roland, Alex (1985). *Model Research: The National Advisory Committee for Aeronautics 1915–1958*, Volume 1. Washington, D.C.: Scientific and Technical Information Office. 3–4.

161 **"aerial construction and navigation"**: Roland, Alex (1985). *Model Research: The National Advisory Committee for Aeronautics 1915–1958*, Volume 1. Washington, D.C.: Scientific and Technical Information Office. 3–4.

161 **"prominent scientists and engineers"**: Roland, Alex (1985). *Model Research: The National Advisory Committee for Aeronautics 1915–1958*, Volume 1. Washington, D.C.: Scientific and Technical Information Office. 3. Also in Gorn, Michael H. (2001). *Expanding the Envelope: Flight Research at NACA and NASA*. Lexington, KY: University Press of Kentucky. 23.

162 **"a post in Gottingen right away"**: Rouse, Hunter and Simon Ince (1957). *History of Hydraulics*, 1st ed. Iowa City: Institute of Hydraulic Research. 230.

162 **"in the field of mechanics"**: Rouse, Hunter and Simon Ince (1957). *History of Hydraulics*, 1st ed. Iowa City: Institute of Hydraulic Research. 231.

162 **"power requirements for flight"**: Based on a newspaper report of the talk reproduced in Lanchester, F. W. (1907). *Aerodynamics*. London: Constable & Co., Ltd.

162 **"a German translation of his *Aerodynamics*"**: Ackroyd, J. A. D. "Lanchester's Aerodynamics." 93. In Fletcher, John (n.d.). *The Lanchester Legacy*. Lanchester Press, Inc.

163 **"benefit there from"**: Fletcher, John (n.d.). *The Lanchester Legacy*. Lanchester Press, Inc. 93.

163 **"had long been taught"**: Fletcher, John (n.d.). *The Lanchester Legacy*. Lanchester Press, Inc. 96.

163   **"return-flow wind tunnel":** Baals, Donald D. and William R. Corliss (1981). *Wind Tunnels of NASA* (NASA SP-440). Washington, D.C.: Scientific and Technical Information Office. 11.

163   **"a university course on 'the mechanics of flight'":** O'Conner, J. J. and Robertson, E. F. (2000). "Richard von Mises." http://www-groups.dcs.st-and.ac.uk/~history/Mathematicians/Mises.html. Accessed November 2004.

164   **"of all time":** Rouse, Hunter and Simon Ince (1957). *History of Hydraulics*, 1st ed. Iowa City: Institute of Hydraulic Research. 223.

164   **"the University of Moscow":** Roland, Alex (1985). *Model Research: The National Advisory Committee for Aeronautics 1915–1958*, Volume 1. Washington, D.C.: Scientific and Technical Information Office. 3.

164   **"for years afterwards":** Baals, Donald D. and William R. Corliss (1981). *Wind Tunnels of NASA* (NASA SP-440). Washington, D.C.: Scientific and Technical Information Office. 10.

164   **"and in Auteuil":** Roland, Alex (1985). *Model Research: The National Advisory Committee for Aeronautics 1915–1958*, Volume 1. Washington, D.C.: Scientific and Technical Information Office. 3.

164   **"concerned with aviation":** Roland, Alex (1985). *Model Research: The National Advisory Committee for Aeronautics 1915–1958*, Volume 1. Washington, D.C.: Scientific and Technical Information Office. 3.

165   **"seemed to be going nowhere":** Roland, Alex (1985). *Model Research: The National Advisory Committee for Aeronautics 1915–1958*, Volume 1. Washington, D.C.: Scientific and Technical Information Office. 10–11.

165   **"located in the Washington Navy Yard":** Anderson, J. D., Jr. (1997). *A History of Aerodynamics*. Cambridge, UK and New York: Cambridge University Press. 299.

166   **"Advisory Committee for Aeronautics":** Anderson, Frank W. Jr. (1976). *Orders of Magnitude: A History of NACA and NASA, 1915–1976* (NASA SP-4403). Washington, D.C.: Scientific and Technical Information Office. 1.

166   **"NASA's early history":** Roland, Alex (1985). *Model Research: The National Advisory Committee for Aeronautics 1915–1958*, Volume 1. Washington, D.C.: Scientific and Technical Information Office.

166   **"stature and prestige":** Tobin, James (2003). *To Conquer the Air: The Wright Brothers and the Great Race for Flight*. New York: Free Press. 4.

167 **"his distinguished reputation":** Crouch, Tom D. (2002). *A Dream of Wings.* New York and London: W.W. Norton. 128.

167 **"the attempted coup":** Tobin, James (2003). *To Conquer the Air: The Wright Brothers and the Great Race for Flight.* New York: Free Press. 200–201.

168 **"reward for their work":** Tobin, James (2003). *To Conquer the Air: The Wright Brothers and the Great Race for Flight.* New York: Free Press. 360–361.

168 **"Or so it seemed":** Roland, Alex (1985). *Model Research: The National Advisory Committee for Aeronautics 1915–1958,* Volume 1. Washington, D.C.: Scientific and Technical Information Office. 1.

168 **"fate had so cruelly withheld":** Heppenheimer, T. A. (2003). *First Flight: The Wright Brothers and the Invention of the Airplane.* New York: John Wiley & Sons, Inc. 330ff.

169 **"around dirigible air hulls":** Baals, Donald D. and William R. Corliss (1981). *Wind Tunnels of NASA* (NASA SP-440). Washington, D.C.: Scientific and Technical Information Office. 10.

169 **"for greater efficiency":** Heppenheimer, T. A. (2003). *First Flight: The Wright Brothers and the Invention of the Airplane.* New York: John Wiley & Sons, Inc. 332–333.

170 **"knew what Curtiss had done":** Heppenheimer, T. A. (2003) *First Flight: The Wright Brothers and the Invention of the Airplane.* New York: John Wiley & Sons, Inc. 334.

171 **"made the general principle rather elusive":** Letter dated November 13, 1915, from Edgar Buckingham to Rayleigh, handwritten on official stationery imprinted "Address reply to Bureau of Standards." 1–2.

172 **"They use Rayleigh's equation for fluid resistance":** Stanton, T. E. and J. R. Pannell (1914). "Similarity of Motion in Relation to the Surface Friction of Fluids." *Proceedings of the Royal Society of London, Series A, Containing Papers of a Mathematical and Physical Character.* Vol. 90, No. 619 (July 1, 1914). 201.

172 **"to yield similar motions":** Stanton, T. E. and J. R. Pannell (1914). "Similarity of Motion in Relation to the Surface Friction of Fluids." *Proceedings of the Royal Society of London, Series A, Containing Papers of a Mathematical and Physical Character.* Vol. 90, No. 619 (July 1, 1914). 214.

172   **"physicists will be sure to read it":** Letter dated November 13, 1915, from Edgar Buckingham to Rayleigh, handwritten on official stationery imprinted "Address reply to Bureau of Standards." 3.

173   **"with temperature and heat":** Buckingham, Edgar (1900). *Outline of a Theory of Thermodynamics*. New York: Macmillan. 16.

173   **"the laws of the logic are the laws of the laws of science":** Frege, Gottlob. Section 87. *The Foundations of Arithmetic*. Translated by J. L. Austin. Second revised edition. Evanston, IL: Northwestern University Press. 1980. 99.

174   **"Hertz proposed his own account":** Hertz, Heinrich (1894, tr. 1899). *The Principles of Mechanics Presented in a New Form*. Reprint available 2004: New York: Dover Publications.

175   **"relevant to flow":** Landa, Edward R. and John R. Nimmo (2003). "The Life and Scientific Contribution of Lyman J. Briggs." *Soil Science Society of America Journal*, Vol. 67, No. 3 (May/June 2003). 681–693; quote is from 685.

175   **"the interstices of soil":** Reynolds, Osborne (1879). "On Certain Dimensional Properties of Matter in the Gaseous State. Part I. Experimental Researches on Thermal Transpiration of Gases through Porous Plates and on the Laws of Transpiration and Impulsion, Including an Experimental Proof that Gas is Not a Continuous Plenum. Part II. On an Extension of the Dynamical Theory of Gas, Which Includes the Stresses, Tangential and Normal, Caused by a Varying Condition of Gas, and Affords an Explanation of the Phenomena of Transpiration and Impulsion." *Philosophical Transactions of the Royal Society of London*, Vol. 170. 321.

176   **"by the porosity of soil":** Buckingham, E. (1904). "Contributions to our knowledge of the aeration of soils." USDA, Bureau of Soil Bulletin No. 25. U.S. Government Printing Office, Washington, D.C. 45.

176   **"Buckingham's show of brilliance":** Buckingham's work on soils is receiving more appreciation as time goes on. In late 2004, we find this description of a talk given at the University of California at Berkeley: "His 1907 publication 'Studies on the Movement of Soil Moisture' provided a dynamical theory of water movement in porous media and unified saturated-unsaturated flow. More general than Darcy's contribution, it preceded M. King Hubbert's Theory of Ground-water Motion by more than three decades. It is likely that Buckingham was the first to recognize the dependence of hydraulic conductivity on capillary potential (widely known as relative

permeability), and thus his work forms part of the foundations of multi-phase flow. A century later, reading Buckingham's paper in original is remarkably educational. Setting out to explain a counterintuitive field observation that soils of arid regions, at depths a little below the surface, are wetter and hold their moisture for longer periods than do soils of humid regions in dry seasons, Buckingham combined simple experiments with superb theoretical insight to revolutionize a new field. Buckingham's approach as a scientist helps us improve the quality of our own science. Exposing students to papers such as Buckingham's must be part of our regular curriculum." From an abstract of a talk, "Buckingham 1907: An Appreciation," by T. Narasimham. Announcement of Talk, Materials Science and Engineering, 2004.

177   **"to look upon them"**: Letter from Whitney to Briggs, August 26, 1905. Quoted in Landa, Edward R. and John R. Nimmo (2003). "The Life and Scientific Contribution of Lyman J. Briggs." *Soil Science Society of America Journal*, Vol. 67, No. 3 (May/June 2003). 686.

177   **"mathematical physic (sic) measurements"**: Letter dated Sept. 2, 1905 from Whitney about reasons for Briggs's transfer. Quoted in Landa, Edward R. and John R. Nimmo (2003). "The Life and Scientific Contribution of Lyman J. Briggs." *Soil Science Society of America Journal*, Vol. 67, No. 3 (May/June 2003).

177   **"a German-language scientific journal"**: Einstein, Albert (1905). "Review of E. Buckingham, 'On Certain Difficulties Which are Encountered in the Study of Thermodynamics.'" (*Philosophical Magazine and Journal of Science* 9 (Series 6): 208–214) *Beiblatter zu den Annalen der Physik* 29: 635.

177   **"his work on foundations of thermodynamics"**: There is a happy ending to this story. Briggs later went to the National Bureau of Standards (NBS), where he worked on wind tunnel design and propeller research, among other things, and eventually became director of the NBS. While Briggs was head of the Engineering Physics division, Buckingham was employed by him as a consultant and was made the first NBS scientist to be granted the status of independent researcher, freeing him of administrative duties and allowing him to pursue research in his first love, thermodynamics. This was from 1923 until Buckingham's retirement at age 70 in 1937.

177   **"any more justification than another, a priori"**: Buckingham, Edgar (1912). "On the Time Scale." *Philosophical Magazine*, April 1912. 651.

178   **"that time 'really' does progress in this way":** Buckingham, Edgar (1912). "On the Time Scale." *Philosophical Magazine*, April 1912. 651.

178   **"when our choice is free":** Buckingham, Edgar (1912). "On the Time Scale." *Philosophical Magazine*, April 1912. 652.

178   **"on the special theory of relativity":** Einstein, Albert (1905). "On the Electrodynamics of Moving Bodies." Published as *Zur Elektrodynamik bewegter Körper* in Annalen der Physik. 17:891, 1905, from the English translation in *The Principle of Relativity*. London: Methuen and Company. 1923.

178   **"independent of the clock itself":** Buckingham, Edgar (1912). "On the Time Scale." *Philosophical Magazine*, April 1912. 652.

178   **"in all branches of physics":** Buckingham, Edgar (1912). "On the Time Scale." *Philosophical Magazine*, April 1912. 652.

179   **"in the form of equations":** Buckingham, Edgar (1912). "On the Time Scale." *Philosophical Magazine*, April 1912. 652.

179   **"to a Latin dictionary":** Buckingham, Edgar (1912). "On the Time Scale." *Philosophical Magazine*, April 1912. 652.

179   **"definite knowledge of physics":** Nichols, Edward Leamington (1904). "The Fundamental Concepts of Physical Science." In Sopka, Katherine P. (ed.). Ibid. 95–96.

182   **"in heat and electromagnetics":** Buckingham, Edgar (1914a). "On Physically Similar Systems." *Journal of the Washington Academy of the Sciences*, July 1914. 353.

182   **"derived from them":** Buckingham, Edgar (1914a). "On Physically Similar Systems." *Journal of the Washington Academy of the Sciences*, July 1914. 348.

183   **"the ideal picture and its real prototype":** Buckingham, Edgar (1921). "Notes on the Theory of Dimensions." *Philosophical Magazine*, v. 42, No. 251, November 1921. 698.

183   **"as additional arguments of the function φ":** Buckingham, Edgar (1914a). "On Physically Similar Systems." *Journal of the Washington Academy of the Sciences*, July 1914. 349.

183 **"the number of fundamental units"**: Buckingham, Edgar (1914a). "On Physically Similar Systems." *Journal of the Washington Academy of the Sciences*, July 1914. (Quote refers to equation (8).)

184 **"of all the quantities involved"**: Buckingham, Edgar (1914a). "On Physically Similar Systems." *Journal of the Washington Academy of the Sciences*, July 1914. 350.

185 **"without dragging in so much mathematical machinery"**: Letter from Edgar Buckingham to Rayleigh dated November 13, 1915, handwritten on official stationery imprinted "Address reply to Bureau of Standards." 1–2.

185 **"therefore be satisfied"**: Buckingham, Edgar (1914a). "On Physically Similar Systems." *Journal of the Washington Academy of the Sciences*, July 1914. 352.

187 **"that holds between quantities"**: Buckingham, Edgar (1914a). "On Physically Similar Systems." *Journal of the Washington Academy of the Sciences*, July 1914. 347.

187 **"Some instead credit Vaschy"**: Vaschy, A. (1892). *Annales télégraphiques*. 25–28, 189–21.

188 **"as the π theorem"**: Buckingham, Edgar (1921). "Notes on the Theory of Dimensions." *Philosophical Magazine*, v. 42, No. 251, November 1921. 696.

188 **"he confined his attention to mechanical quantities"**: Buckingham, Edgar (1921). "Notes on the Theory of Dimensions." *Philosophical Magazine*, v. 42, No. 251, November 1921. 696n.

189 **"In the paper that appeared in October 1914"**: Buckingham, E. (1914b). "On physically similar systems: Illustrations of the use of dimensional equations." *Physical Review*, IV(4):345–376.

189 **"to describe by an equation"**: Buckingham, E. (1914b). "On physically similar systems: Illustrations of the use of dimensional equations." *Physical Review*, IV(4): 345.

189 **"of the remaining Πs"**: Buckingham, E. (1914b). "On physically similar systems: Illustrations of the use of dimensional equations." *Physical Review*, IV(4): 348.

190 **"physical relation in question"**: Buckingham, E. (1914b). "On physically similar systems: Illustrations of the use of dimensional equations." *Physical Review*, IV(4): 352.

192   **"is represented by I²":** Buckingham, E. (1914b). "On physically similar systems: Illustrations of the use of dimensional equations." *Physical Review*, IV(4): 346.

192   **"as exponents that indicate repeated operations":** Juliet Floyd has an in-depth discussion of Wittgenstein's use of "operations" in the *Tractatus*. She points out the distinction he makes between operation and function and cites passages in Whitehead about operations that may have been influential to his thoughts on the topic.

192   **"which produces the thing p":** Quoted in Floyd, Juliet. "Number and Ascriptions of Number in Wittgenstein's *Tractatus*." In *Future Pasts: The Analytic Tradition in Twentieth-Century Philosophy*. New York: Oxford University Press, 2001. 145–191.

194   **"at the same slip ratio":** Buckingham, E. (1914b). "On physically similar systems: Illustrations of the use of dimensional equations." *Physical Review*, IV(4): 370.

195   **"one of the two arguments of φ":** Buckingham, E. (1914b). "On physically similar systems: Illustrations of the use of dimensional equations." *Physical Review*, IV(4): 371.

195   **"much more important":** Buckingham, E. (1914b). "On physically similar systems: Illustrations of the use of dimensional equations." *Physical Review*, IV(4): 371.

195   **"by ordinary methods":** Buckingham, E. (1914b). "On physically similar systems: Illustrations of the use of dimensional equations." *Physical Review*, IV(4): 372.

195   **"based on 'the principle of homogeneity.'":** Buckingham, E. (1914b). "On physically similar systems: Illustrations of the use of dimensional equations." *Physical Review*, IV(4): 356.

196   **"independent of the clock itself":** Buckingham, Edgar (1912). "On the Time Scale." *Philosophical Magazine*, April 1912. 652.

196   **"from the other two":** Buckingham, E. (1914b). "On physically similar systems: Illustrations of the use of dimensional equations." *Physical Review*, IV(4). 372–373.

197 **"the three that physicists ordinarily use"**: Buckingham, E. (1914b). "On physically similar systems: Illustrations of the use of dimensional equations." *Physical Review*, IV(4). 373.

197 **"shall remain invariable"**: Buckingham, E. (1914b). "On physically similar systems: Illustrations of the use of dimensional equations." *Physical Review*, IV(4). 373.

198 **"of no sensible importance"**: Buckingham, E. (1914b). "On physically similar systems: Illustrations of the use of dimensional equations." *Physical Review*, IV(4). 373.

198 **"the problem of the liquid globe"**: Letter from Edgar Buckingham to Rayleigh dated March 1, 1916. Handwritten version. 3.

199 **"fixed by the law of gravitation"**: Buckingham, E. (1914b). "On physically similar systems: Illustrations of the use of dimensional equations." *Physical Review*, IV(4). 375.

199 **"the difference is mainly one of form"**: Letter from Edgar Buckingham to Rayleigh dated November 13, 1915, handwritten on official stationery imprinted "Address reply to Bureau of Standards." 4.

199 **"The paper is summarized"**: Buckingham, E. (1914b). "On physically similar systems: Illustrations of the use of dimensional equations." *Physical Review*, IV(4). 376.

200 **"more than the thought"**: Sterrett, Susan G. (2004). "How Many Thoughts Can Fit in the Form of a Proposition?" http://philsci-archive.pitt.edu/archive/00001816/. Accessed March 15, 2005.

200 **"the symbols of all the quantities Q"**: Buckingham, E. (1914b). "On physically similar systems: Illustrations of the use of dimensional equations." *Physical Review*, IV(4) 376.

202 **"that seems to be engulfing us"**: Buckingham, E. (1914b). "On physically similar systems: Illustrations of the use of dimensional equations." *Physical Review*, IV(4) 376.

203 **"in one another's places"**: Letter from Ludwig Wittgenstein to Bertrand Russell from the Allegasse in Vienna dated January 1913. In *Notebooks: 1914–1916*. Ibid. Appendix III.

203 **"in Vienna in July 1914"**: McGuinness, Brian (1988). Ibid. 264.

204 **"for a majority of physicists!":** Letter from Rayleigh to Edgar Buckingham, handwritten on plain paper, from Terling Place, Wickham [Essex] dated October 30, 1915. 1.

205 **"your opinion on this point":** Letter from Ludwig Wittgenstein to Bertrand Russell from the Skjolden dated January 1914. Tr. by G. E. M. Anscombe. In *Notebooks: 1914–1916*. Ibid. Appendix III.

Chapter 8: A World Made of Facts

207 **"to the end of September 1918":** McGuinness, Brian F. (1988). *Wittgenstein: A Life: Young Ludwig 1889–1921*. Berkeley: University of California Press. 264.

207 **"saved him from suicide during this time":** McGuinness, Brian (1988). Ibid. 264.

207 **"David Pinsent, had by now died":** McGuinness, Brian (1988). Ibid. 264.

207 **"shortly before Wittgenstein was captured":** McGuinness, Brian (1988). Ibid. 268.

208 **"on the manuscript":** McGuinness, Brian (1988). Ibid. 271.

208 **"having books sent to him":** McGuinness, Brian (1988). Ibid. 276.

208 **"I've solved our problems finally":** von Wright, G. H. (n.d.). *Wittgenstein*. Ibid. 72.

208 **"without an academic home there."** The details are not easy to summarize. See G. H. Hardy (1970). *Bertrand Russell and Trinity*. Cambridge, UK: Cambridge University Press.

208 **"whether he is alive or dead":** Russell, Bertrand (1985). *Philosophy of Logical Atomism*. Chicago and LaSalle, Ill.: Open Court. 35. (First published in 1918.)

209 **"have been edited and translated":** Schmitt, Richard Henry (2003). "Frege's Letters to Wittgenstein about the *Tractatus*." With an introduction and translation by Richard Henry Schmitt. (Edited by Allan Janik and Christian Berger.) *Bertrand Russell Society Quarterly*, November 2003.

209 **"expect so much of the reader":** Frege to Wittgenstein, 28 June 1919. In Schmitt, Richard Henry (2003). Ibid. 7.

209 **"where it obtains":** Frege to Wittgenstein, 28 June 1919. In Schmitt, Richard Henry (2003). Ibid. 8. These remarks are not quite as critical of Wittgenstein as they might at first appear, for Frege had already in other essays written about his frustration with this feature of language in general. He had even opined in a 1906 letter to Husserl that the question of whether equipollent propositions are congruent is not likely to be settled (Beaney 1997; 305). Frege's views on this point are discussed in Sterrett (2004), "How Many Thoughts Can Fit in the Form of a Proposition?," cited previously.

209 **"who read and understood it":** Wittgenstein, Ludwig, D. F. Pears and B. F. McGuinness (eds.) (1972). *Tractatus Logico-Philosophicus*, reprinted with corrections. New York, Humanities Press. 3.

210 **"anticipated by someone else":** Wittgenstein, Ludwig, D. F. Pears and B. F. McGuinness (eds.) (1972) *Tractatus Logico-Philosophicus*, reprinted with corrections. New York, Humanities Press. 3.

210 **"for the stimulation of my thoughts":** Wittgenstein, Ludwig, D. F. Pears and B. F. McGuinness (eds.) (1972). *Tractatus Logico-Philosophicus*, reprinted with corrections. New York, Humanities Press. 3. Regarding this comment of Wittgenstein's, some philosophers have read significance into the contrast between Wittgenstein's reference to Russell's works as "the writings of my friend" and Frege's as "Frege's great works," taking the contrast to reflect a great difference in Wittgenstein's regard for their philosophical work. I would caution against making too much of this, keeping in mind that Wittgenstein and Russell had a close friendship in addition to their interactions as philosophers, and they worked as colleagues on problems he referred to as "our" problems. Their relationship had emotional highs and lows. Although Wittgenstein could become emotionally devastated by criticism from Frege, his relationship with Frege was more distant and had more of a professional than personal character.

210 **"the logic of our language is misunderstood":** Wittgenstein, Ludwig, D. F. Pears and B. F. McGuinness (eds.) (1972). *Tractatus Logico-Philosophicus*, reprinted with corrections. New York, Humanities Press. 3.

210 **"but an activity":** Wittgenstein, Ludwig, D. F. Pears and B. F. McGuinness (eds.) (1972). *Tractatus Logico-Philosophicus*, reprinted with corrections. New York, Humanities Press. 4.112.

210   **"pass over in silence":** Wittgenstein, Ludwig, D. F. Pears and B. F. McGuinness (eds.) (1972). *Tractatus Logico-Philosophicus*, reprinted with corrections. New York, Humanities Press. 3.

210   **"questions about logic":** Letter from Frege to Husserl in 1906. In Beany, Michael (1997). *A Frege Reader.* Oxford: Blackwell Publishers. 303.

211   **"the 'logic' of jazz":** Janik, Allan and Stephen Toulmin (1973). *Wittgenstein's Vienna.* New York: Simon and Schuster. 175.

211   **"Weiniger's influence in particular":** Monk, Ray (1990). *Wittgenstein: The Duty of Genius.* New York: Penguin Books. 20–25.

212   **"in a Paris courtroom":** von Wright, G. H. (n.d.). *Wittgenstein.* Ibid. 72.

212   **"of apparently complex things":** Russell, Bertrand (1985). *Philosophy of Logical Atomism.* Chicago and LaSalle, Ill.: Open Court. 52. (First published in 1918.)

212   **"between name and thing named":** Russell, Bertrand (1985). Ibid. 46.

213   **"just a symbol":** Russell, Bertrand (1985). Ibid. 44.

213   **"the facts they stand for are complex":** Russell, Bertrand (1985). Ibid. 52.

213   **"his masterwork, *Principia Mathematica*":** Russell, Bertrand (1985). Ibid. 59–60.

213   **"nor therefore of symbolism":** Russell, Bertrand (1985). Ibid. 45.

213   **"In his essay 'Pictures and Form'":** McGuinness, Brian (2002). "Pictures and Form" in McGuinness, Brian (2002). *Approaches to Wittgenstein: Collected Papers.* London and New York: Routledge. 61–81.

214   **"and so on":** Wittgenstein, Ludwig. *Tractatus Logico-Philosophicus.* London: Routledge Kegan. 1927. 7.

214   **"that any truth-function must have (5's)":** McGuinness, Brian (2002). "Pictures and Form" in McGuinness, Brian (2002). Ibid. 61.

214   **"of what it is to be a proposition":** McGuinness, Brian (2002). "Pictures and Form" in McGuinness, Brian (2002). Ibid. 65–66.

214   **"throughout the *Tractatus*":** McGuinness, Brian (2002). "Pictures and Form" in McGuinness, Brian (2002). Ibid. 80.

215   **"the general form of a proposition"**: Quoted in and translated by
        McGuinness, Brian (2002). "Pictures and Form" in McGuinness, Brian
        (2002). Ibid. 79.

217   **"the language of gramophone records"**: Wittgenstein, Ludwig, D. F. Pears
        and B. F. McGuinness (eds.) (1972). *Tractatus Logico-Philosophicus*,
        reprinted with corrections. New York: Humanities Press. 4.0141.

217   **"the language of the gramophone record"**: Wittgenstein, Ludwig
        (1971/1922). *Tractatus Logico-Philosophicus*, edited by C. K. Ogden. London:
        Routledge & Kegan Paul, Ltd. 4.014.

222   **"by means of dolls, etc."**: Wittgenstein, Ludwig, G. H. von Wright and G. E.
        M. Anscombe (eds.) (1984). *Notebooks: 1914–1916*, 2nd ed. Chicago,
        University of Chicago Press. 7e.

224   **"what he did not say in it"**: McGuinness, Brian (1988). Ibid. 288.

224   **"by being silent about it"**: Letter from Ludwig Wittgenstein to Ludwig von
        Ficker. Quoted in McGuinness, Brian (1988). Ibid. 288.

226   **"also the general form of a proposition"**: Wittgenstein, Ludwig, D. F. Pears
        and B. F. McGuinness (eds.) (1972). *Tractatus Logico-Philosophicus*,
        reprinted with corrections. New York, Humanities Press. xv.

226   **"of the (complete physical) empirical equation"**: Russell also goes on to say
        that the symbol describes a process by which, given the atomic propositions,
        all other propositions can be formed. I think it dubious that Wittgenstein
        would have endorsed putting it this way, as the atomic propositions are the
        result of an analysis, not a starting point, so I have not included that part of
        the quote.

229   **"'The object is simple *for me!*'"**: Wittgenstein, Ludwig, G. H. von Wright
        and G. E. M. Anscombe (eds.) (1984). *Notebooks: 1914–1916*, 2nd ed.
        Chicago, University of Chicago Press. 70.

238   **"the ratio of inertial force to viscous force"**: Becker, H. A. (1976).
        Dimensionless Parameters: Theory and Methodology. Applied Science
        Publishers. 43–46.

240   **"Henk Visser has pointed out"**: Visser, Henk (1999). "Boltzmann and
        Wittgenstein, or How Pictures Became Linguistic." *Synthese*, v. 119 135–156.

240   **"in the place of real things"**: Ostwald, Wilhelm (1902). Vorlesen uber
        Naturphilosophie. Leipzig: Veit. 100. Quoted in Visser, Henk (1999). Ibid.

242 **"It's a reductio ad absurdum":** Ian Proops also describes Wittgenstein's rea-
soning here as a reductio ad absurdum argument, but a very different one.
Proops, Ian (2004). "Wittgenstein's Logical Atomism." Proops (*Stanford
Encyclopedia of Philosophy*, November 2004).

245 **"As one scholar recently put it":** In the appendix to his *Logic and Language
in Wittgenstein's Tractatus*, Proops shows the original ordering of statements
in an earlier draft of the *Tractatus*, called the *Prototractatus*, which shows
that the anaphoric reference of "this" in 4.02—"This we see from the fact
that we understand the sense of the propositional sign without having had it
explained to us."—is "A proposition is a picture of reality." Proops, Ian
(2000). *Logic and Language in Wittgenstein's Tractatus*. Garland Publishing.
104–105.

245 **"what each word stands for":** Specifically, in TLP 4.022: "Man possesses the
ability to construct languages capable of expressing every sense, without
having any idea how each word has meaning or what its meaning is—just as
people speak without knowing how the individual sounds are produced." In
Wittgenstein, Ludwig. D. F. Pears and B. F. McGuinness (eds.) (1972).
Tractatus Logico-Philosophicus. Reprinted with corrections. New York:
Humanities Press. 19.

249 **"without having had it explained to us":** Proops, Ian (2000). *Logic and
Language in Wittgenstein's Tractatus*. New York: Garland Publishing. 104-
105. (The phrase quoted is TLP 4.02, in Wittgenstein, Ludwig, Ibid. 21.

250 **"he had already grasped":** McGuinness, B. F. "The Grundgedanke of the
Tractatus." In McGuinness, Brian (2001). *Approaches to Wittgenstein:
Collected Papers*. New York and London: Routledge. 111.

251 **"causing people to misunderstand his theory":** Sterrett, Susan G. (2002).
"Darwin's Analogy Between Artificial and Natural Selection: How Does It
Go?" *Studies in History and Philosophy of Science*, Part C. 33.

Acknowledgments

The thesis that propositions and facts in Wittgenstein's *Tractatus* are like models and dimensionless parameters came to me about a dozen years ago. The first person to whom I showed my short writeup of it was the philosopher John McDowell—in spite of my suspicion that, although Wittgenstein's philosophy was one of his areas of expertise, the style of philosophy might not be to his taste. The second person I showed my short writeup to was the philosopher of science Clark Glymour—in spite of my suspicion that, although this style of philosophy was one of his areas of expertise, Wittgenstein's philosophy might not be to his taste. There was some truth in these suspicions, yet both were very encouraging, and their encouragement when I was at my most tentative was important.

Then came the hard work. The long task of exploring territory on Wittgenstein previously covered by others was made much easier by the many published biographies and memoirs of Wittgenstein. Four deserve special acknowledgement here for their value in writing this book: von Wright's *Wittgenstein*, Toulmin and Janik's *Wittgenstein's Vienna*, Brian McGuinness' *Wittgenstein: A Life* (recently reissued as *Young Ludwig*), and Ray Monk's *Wittgenstein: The Duty of Genius*.

I presented the idea publicly for the first time at a conference in Vienna at the turn of the century: HOPOS 2000 (History of Philosophy of Science), held July 5, 2000. Comments from audience members there helped me see how to sharpen the ideas, and its selection for publication in the Vienna Circle Institute yearbook was further encouragement. The encouragement of many others who knew of my book project helped me stick with it: Brian McGuinness, Allan Janik, Michael Heidelberger, Ian Proops, Juliet Floyd, Jim Bogen, and Michael Biggs. In November of 2000, I gave a longer version at UNC-Chapel Hill in North Carolina, and received more helpful comments, particularly from Jay Rosenberg. The British philosopher Simon Blackburn gently pointed out to me that my reference to Edgar Buckingham as Lord Buckingham had to be in error, which made me soberly realize I had more historical territory yet to survey. (Many of the institutions in the U.S. had been modeled on British ones and had the same or similar names, leading to confusion and making research more difficult.)

There was practically nothing published about Edgar Buckingham's life. Lisa Greenhouse at the National Institute of Standards and Technology library and Marjorie Ciarlante at the National Archives in Washington, D.C., gave generously of their time to help me locate what was available in their institutions. Alex Roland, an ex-NASA historian and the author of *Model Research: The National Advisory Committee for Aeronautics 1915–1958*, gave me feedback about what I planned to say about Buckingham and the early days of the U.S. ACA/NACA. The historian Malachi Hacohen, author of a definitive biography of Karl Popper's early years, read parts of the manuscript, too, and provided encouragement and research tips. Allan Janik answered my questions about Wittgenstein's correspondence. The philosopher of science Marc Lange, himself intrigued by dimensional analysis and engineering scale models in the history of ship design, read and commented on the entire manuscript, for which I am most grateful. Peter Spirites read—and reread rewrites of—technical sections of the book whenever I asked.

Stephen Morrow of Pi Press was unfailingly upbeat about the prospects of turning my tentative idea into a book that would interest the general reader as well as professional philosophers. A fellowship from the Woodrow Wilson Foundation and a generous paid junior leave from Duke University gave me a much-needed year to complete the book.

Index

Page numbers followed by *n* signify endnotes.